Conference Board of the Mathematical Sciences

REGIONAL CONFERENCE SERIES IN MATHEMATICS

supported by the
National Science Foundation

Number 47

TOPICS IN DYNAMIC BIFURCATION THEORY

by

JACK K. HALE

Published for the
Conference Board of the Mathematical Sciences
by the
American Mathematical Society
Providence, Rhode Island

Expository Lectures
from the CBMS Regional Conference
held at the University of Texas at Arlington
June 16–20, 1980

This monograph was written while the author held a Guggenheim Fellowship. Parts of the research efforts reported herein were supported by the United States Air Force Office of Scientific Research under AFOSR 76-3092C, in part by the United States Army Research Office under ARO-DAAG-29-79-C-0161 and in part by the National Science Foundation under MCS 79-05774-02.

1980 *Mathematics Subject Classification.* Primary 34C15, 34C28, 34C35, 34D30; Secondary 35B32, 35B35, 35B40, 58F13, 58F14, 58F10.

Library of Congress Cataloging in Publication Data

Hale, Jack K.

Topics in dynamic bifurcation theory.

(Regional conference series in mathematics; no. 47)

"Expository lectures from the CBMS regional conference, held at the University of Texas at Arlington, June 16–20, 1980."

Bibliography: p.

1. Differential equations. 2. Nonlinear oscillations. 3. Bifurcation theory. I. Conference Board of the Mathematical Sciences. II. Title. III. Series.

QA1.R33 no. 47 [QA372] 515.3′5 81-3445

ISBN 0-8218-1698-5 AACR2

CONTENTS

To Lamberto Cesari on his 70th Birthday

1. Introduction

My objective in this paper is to give some of the basic results in the theory of bifurcation in differential equations. It is difficult to trace the historical development of any important concept and bifurcation theory is no exception. However, a careful study of literature shows that Poincaré [1], [2] and Liapunov [1], [2] are responsible for our present basic philosophy as well as several of the fundamental ideas of the methods that we presently employ. These two persons can be directly linked with the importance of exchanges of stability, the occurrence of complicated motions in dynamical systems, the principle of reduction to lower dimensional problems, the philosophy of genericity and the transformation theory of differential equations that is so important in obtaining approximations of the center manifold and the flow on the center manifold. In many respects, we are still exploiting the ideas of these two giants.

After the initial impetus of Poincaré and Liapunov, it is somewhat surprising that modern bifurcation theory did not appear at an earlier date. It is perhaps true that the ideas of Liapunov connected with bifurcation theory were being developed more extensively than the corresponding ones of Poincaré. There was a very active group in the U.S.S.R. (consisting of Andronov, Vitt, Khaikin, Bogoliubov, Krylov, Leontovich, Malkin and others) working on critical cases in stability theory, nonlinear oscillations and the general theory of integral manifolds. The techniques developed from the study of these areas are fundamental ingredients in dynamic bifurcation theory (see Hale [1] for some references).

A fundamental step towards modern bifurcation theory in differential equations occurred with the definition of structural stability of Andronov and Pontrjagin [1] in 1937 and the classification of structurally stable systems in the plane. With these concepts, Andronov and Leontovich [1] were able to make precise definitions of types of bifurcation points which had the possibility of being classified. These results were applied extensively to the theory of nonlinear oscillations by Andronov, Vitt and Khaikin in 1937 (the second edition of this book is Andronov, Vitt and Khaikin [1]). As Minorsky [1] said in 1962: "Having established the initial advance in the field of nonlinear oscillations (up to 1940), the Russian scientists maintain their leadership and initiative characterized by a remarkable coordination of efforts between the mathematical and experimental parts of these fundamental researches." There were several important developments in this intervening period by Levinson [1], [2], Cartwright and Littlewood (see Cartwright [1]) on the forced van der Pol and

Lienard equations. However, in western Europe and the United States, the interest in this aspect of differential equations had never been very extensive. In addition, there was little awareness of the developments that had been made in the U.S.S.R., and, as a consequence, some duplication of effort occurred.

Since 1960, there have been extensive developments in the abstract theory of dynamical systems. At the same time, some of this theory has been applied to very interesting problems in the biological and physical sciences. In an attempt to explain phenomena that occur in nature, it has been necessary for researchers to discuss the dynamic bifurcation of specific types of equations in great detail. This has led to an exciting interaction between analytical and theoretical methods.

In this paper, we present some of the concepts and results that play an important role in these areas. When the dimension of the system is one or two, one can obtain a rather complete theory at least for general one parameter families of vector fields. For either several parameter problems or for the dimension of the system greater than two, only partial results are known. On the other hand, the results in low dimension are applicable to higher dimensional problems (even infinite dimensional ones) when the discussion is restricted to a neighborhood of an equilibrium point for which the theory of center manifolds can be employed.

The table of contents expresses in general terms the substance of this paper. The first eight sections deal with structural stability and bifurcation in the low dimensional problems mentioned above. §9 is devoted to the formulation of some of the basic problems in the qualitative theory for a special class of dynamical systems in infinite dimensions. This class is general enough to include many functional differential equations and partial differential equations. §10 is concerned primarily with some types of bifurcation that occur because the base space is infinite dimensional. Due to space, very few proofs are given. Also, there are several important omissions of topics from differential equations that are systematically used throughout, but which are not as well known as they should be. Most notable among these are the general theory of integral manifolds (for general references, see Hale [1]), the center manifold theorem (see Kelley [1]), the theory of transformation to normal forms (for references, see Bibikov [1], Br′juno [1], Henrard [1]) and the general method of averaging (for references, see Hale [1]).

The author is endebted to many colleagues and students whose ideas have been incorporated into these notes–too many to mention by name. He also acknowledges the initiative of Professor Laksmikantham in proposing the CBMS Regional Conference. Finally, Sandra Spinacci has exhibited her usual patience and understanding in the preparation of the final manuscript.

2. On the definition of bifurcation

Suppose X, Z are topological spaces, $U \subset X$ is open, Λ is an open set in a topological space and $f\colon U \times \Lambda \longrightarrow Z$ is a given continuous function. Let

$$S = \{(x, \lambda) \in U \times \Lambda \colon f(x, \lambda) = 0\}$$

be the set of solutions of the equation $f(x, \lambda) = 0$. For a fixed λ, let

$$S_\lambda = \{x\colon (x, \lambda) \in S\}$$

be the "cross-section" of the solution set at λ.

The basic problem is to discuss the dependence of the set S_λ on λ. In a specific problem, one has a prescription which compares S_λ with S_μ for different λ and μ. This comparison is usually made by means of an equivalence relation which divides the sets $\{S_\lambda, \lambda \in \Lambda\}$ into equivalence classes. Given the function f and an equivalence relation $\sim$, we say λ_0 is a *bifurcation point for* $(f, \sim)$ if, for any neighborhood V of λ_0, there are $\lambda_1, \lambda_2 \in V$ such that $S_{\lambda_1} \not\sim S_{\lambda_2}$. This definition is less general than the one in Marsden [1].

A special case is when the equivalence relation specifies that $S_\lambda \sim S_\mu$ if the sets S_λ and S_μ are homeomorphic. This is a very convenient choice when studying the change in the structure of the set of equilibrium points in a differential equation as parameters are varied. In this case, the function f represents the vector field in a differential equation $dx/dt = f(x, \lambda)$. It is also appropriate in differential equations for the study of the set of solutions of some prescribed type; for example, periodic solutions, invariant tori, etc. In this latter case, the topological spaces are defined so that they include only those functions which exhibit this prescribed behavior and the function f could be the differential operator, $f(x, \lambda) = dx/dt - g(x, \lambda)$.

To study more general bifurcations in differential equations, the equivalence defined by homeomorphism is not sufficient. Consider a differential equation $du/dt - g(u, \lambda) = 0$ where $(u, \lambda) \in \Omega \times \Lambda$ and Ω is an open set in some Banach space E. For X, Z Banach spaces of functions from $[0, \infty)$ to E, let $U \subset X$ be defined by $U = \{u \in X\colon u(t) \in \Omega, t \in [0, \infty)\}$. The above equation can be written formally as $f(u, \lambda) = 0$ where $f\colon U \times \Lambda \longrightarrow Z$, $f(u, \lambda)(t) = du(t)/dt - g(u(t), \lambda)$. Assuming everything can be made rigorous and that all solutions are obtained in this way, a comparison of the corresponding sets S_λ and S_μ by homeomorphism will not be very interesting. Thus an alternative approach must be taken.

Suppose the differential equation generates a strongly continuous semigroup $T_\lambda(t)$, $t \geqslant 0$, on Ω. A frequently used concept of equivalence in differential equations is to say that $g(\cdot, \lambda) \sim g(\cdot, \mu)$ if there is a homeomorphism $h: \Omega \longrightarrow \Omega$ such that h maps orbits of $T_\lambda(t)$ onto orbits of $T_\mu(t)$ preserving the sense of direction in time. A vector field $g(\cdot, \lambda_0)$ is structurally stable if there is a neighborhood V of λ_0 such that $g(\cdot, \lambda) \sim g(\cdot, \lambda_0)$ for all $\lambda \in V$. Thus, λ_0 is a *bifurcation point* if λ_0 is not structurally stable.

A different but equivalent formulation of the above concept of equivalence in differential equations was introduced by Andronov and Pontrjagin [1] in 1937 for differential equations in the plane. They gave a characterization of the structurally stable vector fields which will be discussed later. Peixoto [1] generalized these results to arbitrary compact two dimensional manifolds and proved the set of structurally stable vector fields is open and dense. For some time, it was the feeling that this same property should hold true for arbitrary systems. Unfortunately, it was shown by Smale [1] that structurally stable vector fields are not dense in dimension $\geqslant 4$. Williams [1] proved the same result for $n \geqslant 3$. Since many vector fields cannot be compared by this equivalence relation, it becomes necessary to weaken the concept of equivalence. Each new definition of equivalence leads to a new type of stable vector field (ones which are equivalent to everyone in a neighborhood of it) and thus a new type of bifurcation. The ultimate goal is to have the definition restrictive enough to permit classification of the stable ones, but, at the same time, to have the stable vector fields generic; that is, the intersection of a countable sequence of open dense sets. Much of the research in finite dimensional abstract dynamical systems in the last twenty years has been devoted to this general problem. Relevant references are Smale [2], Peixoto [2], [3], Palis and Melo [1], Newhouse [1], Nitecki [1], Shub [1], Guckenheimer [1], Arnol'd [1]. In the next section, we give more specific details.

When the evolutionary equation is infinite dimensional, several new problems arise. This case will be discussed in a later section.

3. Structural stability and generic properties in $\mathbf{R}^n$

Suppose Ω is an open set in $\mathbf{R}^n$ with $\partial\Omega = \Gamma$, $\bar{\Omega} = \Omega \cup \Gamma$.

The space $C^r(\bar{\Omega}, \mathbf{R}^n)$ is the Banach space of functions bounded and continuous together with all derivatives up through order $r \geq 0$ with the norm of f in $C^r(\bar{\Omega}, \mathbf{R}^n)$ being given by the maximum of the supremum over $\bar{\Omega}$ of the norm of f and its derivatives up through order r. Let $\mathcal{X}_n^r = \mathcal{X}_n^r(\bar{\Omega})$ be the set of elements of $C^r(\Omega, \mathbf{R}^n)$ which are transversal to Γ. For any $f \in \mathcal{X}_n^r$, $r \geq 1$, the differential equation

$$\dot{x} = f(x) \tag{3.1}$$

defines a family of transformations $T_f(t)$ on $\bar{\Omega}$ satisfying the semigroup property with $T_f(t)x_0 = x(t, x_0)$, where $x(t, x_0)$ is the solution of (3.1) with $x(0, x_0) = x_0$. Furthermore, for each $x_0 \in \bar{\Omega}$, there are an $\alpha_{x_0} \leq 0$, $\beta_{x_0} \geq 0$, such that the maximal interval of definition of $T_f(t)x_0$ is $[\alpha_{x_0}, \beta_{x_0})$. The number α_{x_0} is either that value α_{x_0} where $T_f(\alpha_{x_0})x_0 \in \partial\Omega = \Gamma$ or $-\infty$ and in this case the interval $[\alpha_{x_0}, \beta_{x_0})$ is $(-\infty, \beta_{x_0})$. The number β_{x_0} is defined in a similar way in the positive direction. The operator $T_f(t)$ on $\bar{\Omega}$ satisfies $T_f(0) = I$, the identity, $T_f(t + s)x = T_f(t)T_f(s)x$ for those t, s for which it is meaningful and $T_f(t)x$ has continuous derivatives up through order r in t, x.

The *orbit* $\gamma_f(x)$ *of* f *through* x is

$$\gamma_f(x) = \bigcup\{T_f(t)x, t \in [\alpha_x, \beta_x]\}.$$

The *ω-limit set* $\omega_f(x)$ and *α-limit set* $\alpha_f(x)$ *of the orbit* $\gamma_f(x)$ are defined by

$$\omega_f(x) = \bigcap_{\tau \geq 0} \text{cl} \bigcup_{t \geq \tau} T_f(t)x, \qquad \alpha_f(x) = \bigcap_{\tau \leq 0} \text{cl} \bigcup_{t \leq \tau} T_f(t)x.$$

An *equilibrium point* or *critical point* of f is a zero of f. A *periodic orbit* of f is an orbit which is a closed curve. A set $M \subset \bar{\Omega}$ is *invariant* if, for each $x \in M$, $T_f(t)x$ is defined for $t \in (-\infty, \infty)$ and belongs to M for $t \in (-\infty, \infty)$. This implies $T_f(t)M = M$ for $t \in (-\infty, \infty)$.

The vector fields in (3.1) are chosen from $\mathcal{X}_n^r$; that is, are transversal to Γ, in order to eliminate technical difficulties with points of contact on Γ. We are discussing the vector fields in $\mathbf{R}^n$, but many of the remarks hold for vector fields on compact manifolds M.

DEFINITION 3.1. Two vector fields f, g in $\mathcal{X}_n^r$, $r \geq 1$, are *equivalent*, $f \sim g$, if there is a homeomorphism $h: \bar{\Omega} \longrightarrow \bar{\Omega}$ such that h maps the orbits defined by f homeomorphically onto the orbits defined by g with the sense of direction in time preserved. An $f \in \mathcal{X}_n^r$

is said to be *structurally stable* if there is a neighborhood U of f such that $f \sim g$ for every $g \in U$. An $f \in X_n^r$ is a *bifurcation point* if f is not structurally stable.

Two important remarks need to be made about this definition. Definition 3.1 would not be meaningful without the condition $r \geqslant 1$. In fact, for $r = 0$, given any vector field f that has an isolated zero at x_0 and any $\epsilon > 0$, there are a $\delta > 0$ and a function g such that $|f - g| < \epsilon$ and $g(x) = 0$ for $|x - x_0| < \delta$. Therefore, no f with an isolated zero could be structurally stable.

In Definition 3.1, it is tempting to require that the mapping h be a diffeomorphism. However, if $f(0) = 0$, $g(0) = 0$, $\partial f(0)/\partial x = A$, $\partial g(0)/\partial x = B$, and $f \sim g$ in a neighborhood of zero, then one can show (see Peixoto [2], [4]) that the eigenvalues of A and B must be proportional. Since one can always make a small perturbation that will change one eigenvalue of A and not the other, it follows that no vector field with a zero could be structurally stable. Thus, the Definition 3.1 would have little meaning. If x_0 is a critical point of f and $A = \partial f(x_0)/\partial x$, then x_0 is said to be *hyperbolic* if the real parts of the eigenvalues of A have nonzero real parts. The point x_0 is a *saddle point of order* k, if it is hyperbolic and there are k eigenvalues of A with positive real parts. The term *saddle point* without the designation of the order will refer to any saddle point of order k with $k \neq 0$ or n.

If $n = 2$, a saddle point of order 1 corresponds to the usual definition of saddle point. For $n = 2$, a saddle point of order 0 or 2 corresponds to a node or focus depending upon whether the eigenvalues of A are real or complex.

If γ is a *periodic orbit* of f, then one can define a Poincaré map near γ in the following way. For any arc C transversal to γ at p_0 and any $p \in C$ sufficiently near p_0, there is a unique $\tau(p) > 0$ such that $T_f(\tau(p))p \in C$, $T_f(t)p \notin C$ for $0 < t < \tau(p)$. The map $p \mapsto T_f(\tau(p))p$ is called the Poincaré map $\pi(p)$. This map in C^r and $\pi(p_0) = p_0$. The periodic orbit γ is *hyperbolic* if no eigenvalue of $\partial\pi(p_0)/\partial p$ has modulus one.

If $n = 2$, the periodic orbit γ is hyperbolic if $d\pi(p_0)/dp \neq 1$. It is instructive to give an equivalent definition in terms of the vector field itself. If $\gamma = \{\phi(t), t \in \mathbf{R}\}$ where $\phi(t)$ is periodic of least period ω and $\dot\phi(t) = f(\phi(t))$ then the *linear variational equation* for ϕ is

$$\dot y = A(t)y, \qquad A(t) = \partial f(\phi(t))/\partial x. \tag{3.2}$$

One characteristic multiplier of this ω-periodic system is 1 since $\dot\phi$ satisfies (3.2). If $X(t)$ is a principal matrix solution of (3.2), then the product of the multipliers is equal to det $X(\omega)$. Thus, if $\rho_\gamma = \exp \omega\sigma_\gamma$, σ_γ real, is the other multiplier, then

$$\sigma_\gamma = \frac{1}{\omega}\int_0^\omega \operatorname{tr} A(s)\,ds. \tag{3.3}$$

One can then easily show that γ is hyperbolic if and only if $\sigma_\gamma \neq 0$, unstable if $\sigma_\gamma > 0$ and asymptotically orbitally stable if $\sigma_\gamma < 0$.

For two dimensional systems, the following result of Andronov and Pontrjagin [1], Peixoto [1], completely solves the problem of structural stability in X_2^r.

THEOREM 3.2. *If $\Sigma_2^r \subset X_2^r$, $r \geq 1$, is the set of structurally stable vector fields in X_2^r, then $f \in \Sigma_2^r$ if and only if the following conditions are satisfied*:

(i) *The critical points of f are hyperbolic.*

(ii) *The periodic orbits of f are hyperbolic.*

(iii) *There is no orbit of f with both the α- and ω-limit sets being saddle points.*

Furthermore, Σ_2^r is open and dense in X_2^r.

The fact that a structurally stable vector field must satisfy (i)–(iii) is very easy to prove. However, the converse is more difficult and relies heavily upon the following result of Hartman [1], [3] Grobman [1], and its extension to diffeomorphisms which is valid in the space of n-dimensional vector fields X_n^r.

THEOREM 3.3 (HARTMAN-GROBMAN). *If $f \in X_n^r$, $r \geq 1, f(x_0) = 0$, and the eigenvalues of $A = \partial f(x_0)/\partial x$ have nonzero real parts then, in a neighborhood of x_0, f is equivalent to the linear equation $\dot{x} = Ax$.*

In Theorem 3.2, the fact that X_2^r is open follows from the definition and the fact that it is dense follows from an argument in transversality theory. See Peixoto [1] for a complete proof.

Condition (i), the Implicit Function Theorem and the compactness of $\bar{\Omega}$ imply that $f \in \Sigma_2^r$ has only a finite number of critical points. Using (ii), (iii) and similar arguments, one shows there is only a finite number of periodic orbits.

The simplicity of the description of the structurally stable systems in two dimensions given by Theorem 3.2 permits a complete classification in terms of certain distinguished graphs (see Peixoto [4]).

To what extent does Theorem 3.2 hold in dimension $n \geq 3$? As remarked earlier, the *structurally stable systems are not dense in X_n^r for $n \geq 3$*. This was proved by Smale [1] for $n \geq 4$ and by Williams [1] for $n \geq 3$. However, there are structurally stable systems in every dimension and on every type of n-dimensional manifold.

Even though Σ_n^r is not dense, it is very important to classify structurally stable vector fields and to find "simple" classes of vector fields which are generic. Let us turn first to the problem of genericity.

The concepts (i), (ii) in Theorem 2 have meaning in $\mathbf{R}^n$. Also, (iii) can be extended in the following way. For any hyperbolic critical point or periodic orbit of a vector field $f \in X_n^r$, one can define the global stable and unstable manifolds in the following way. The stable (unstable) manifold of a hyperbolic critical point x_0 is the set of $x \in \bar{\Omega}$ such that $T_f(t)x \longrightarrow x_0$ as $t \longrightarrow +\infty$ $(-\infty)$. Similar definitions are given for a periodic orbit.

In $\mathbf{R}^2$, condition (iii) is then equivalent to the statement that the stable and unstable manifolds of all critical points and periodic orbits intersect transversally. One can then ask if the vector fields in X_n^r which satisfy these properties are generic in X_n^r. The answer is yes and is the famous theorem of Kupka [1] and Smale [3].

THEOREM 3.4 (KUPKA-SMALE). *The set of vector fields in $\mathcal{X}^r_n$ for which the critical points and periodic orbits are hyperbolic with stable and unstable manifolds intersecting transversally is generic.*

Any vector field satisfying the conditions of Theorem 3.4 will be called a Kupka-Smale (KS) vector field. They can have only a finite number of critical points with the proof being the same as in two dimensions. However, in contrast to two dimensions, there can be an infinite number of periodic orbits if the dimension is $\geqslant 3$ (for an example, see Nitecki [1], Palis and deMelo [1]).

The KS vector fields are dense, but all KS vector fields cannot be structurally stable since the structurally stable systems are not dense in dimension $\geqslant 3$. To find a subset of the KS vector fields which are structurally stable, one must put some further restrictions on the behavior of the α- and ω-limit sets of orbits.

For $f \in \mathcal{X}^r_n$, let

$$L_\alpha(f) = \{p\colon p \in \alpha(q) \text{ for some } q\}, \qquad L_\omega(f) = \{p\colon p \in \omega(q) \text{ for some } q\}.$$

DEFINITION 3.5. Suppose $f \in \mathcal{X}^r_n$. A point $p \in \Omega$ is a *wandering point* of f if there are a neighborhood V of p and $t_0 > 0$ such that if $|t| > t_0$, then $T_f(t)V \cap V = \emptyset$. In the contrary case, p is a *nonwandering point of f*. The *set of nonwandering points of f* is denoted by $\Omega(f)$.

In Definition 3.5, the notation $|t| > t_0$ means for all $t \geqslant t_0$ and all $t < -t_0$ as long as the orbit is defined.

We remark that $\Omega(f) \supset L_\alpha(f) \cup L_\omega(f)$, but it is easy to construct examples for which the inclusion is proper (see, for example, Palis and deMelo [1]).

DEFINITION 3.6. A vector $f \in \mathcal{X}^r_n$ is *Morse-Smale (MS)* if it is KS with a finite number of critical points and periodic orbits with $\Omega(f)$ equal to the set of critical points and periodic orbits.

Some of the basic results on Morse-Smale systems are due to Smale [4], Palis [1] and Palis and Smale [1]. They are summarized in the following theorem which is also valid for vector fields on any compact manifold.

THEOREM 3.7. (1) *The set of MS systems is open and nonempty in $\mathcal{X}^r_n$ for any n.*

(2) *Any $f \in MS$ is structurally stable.*

(3) *The set of gradient vector fields which are MS is open and dense in the set of all gradient vector fields.*

Since the MS systems are structurally stable, they cannot be dense in dimension $n \geqslant 3$. On the other hand, one can ask if there are any other structurally stable systems which are not MS. One way to answer this question is to construct a structurally stable system with infinitely many periodic orbits.

To see how such a situation might arise, suppose $\Omega \subset \mathbf{R}^3$ and $f \in \mathcal{X}^r_3(\bar{\Omega})$ has a hyperbolic periodic orbit γ. Let $W^s(\gamma)$, $W^u(\gamma)$ be the stable and unstable manifolds for γ and let π be a Poincaré map of some transversal r of γ at p and $W^s_r(\gamma) = W^s(\gamma) \cap r$, $W^u_r(\gamma) = W^u(\gamma) \cap r$;

that is, that part of the stable and unstable manifolds in the transversal r. Then $W_r^s(\gamma)$, $W_r^u(\gamma)$ are the local stable and unstable manifolds of the point p as a fixed point of the diffeomorphism π. There is the possibility that $W_r^s(\gamma) \cap W_r^u(\gamma)$ contains points other than the fixed point p of π. Any such point q is called *homoclinic* to p. A point q is called *transverse homoclinic* to p if $W_r^s(\gamma)$ is transversal to $W_r^u(\gamma)$ at q. If q is *transverse homoclinic* to p, then the behavior of the stable and unstable manifold is very bad. In fact, since $\pi W_r^s(\gamma) \subset W_r^s(\gamma)$, $\pi W_r^u(\gamma) \subset W_r^u(\gamma)$ and $q \in W_r^s(\gamma) \cap W_r^u(\gamma)$, $q \neq p$, we must have $\pi^n q \in W_r^s(\gamma) \cap W_r^u(\gamma)$ for all $n = 0, \pm 1, \pm 2, \ldots$ and $\pi^n q \longrightarrow p$ as $n \longrightarrow \infty$. If, in addition, q is transverse homoclinic to p, continuity of the map π implies that the picture near p must be something like the one in Figure 1. The arrows do not represent the direction of a flow as for vector fields, but only that points move in the direction indicated under iterates of π. In Figure 1, we have only indicated some of the complications that are arising from looking at the forward evolution of the unstable manifold. The same type of thing must occur with the stable manifold. Note that there will be infinitely many transverse intersections in any neighborhood of the homoclinic point q.

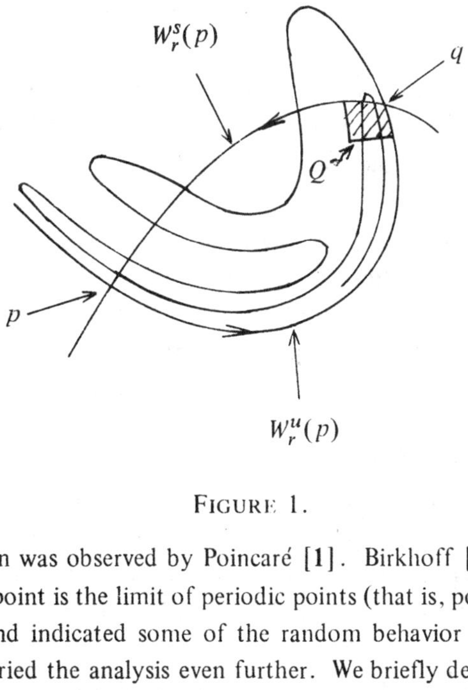

FIGURE 1.

This phenomenon was observed by Poincaré [1]. Birkhoff [1] proved that every transverse homoclinic point is the limit of periodic points (that is, points x such that $\pi^n x = x$ for some integer n) and indicated some of the random behavior that occurs near these points. Smale [1] carried the analysis even further. We briefly describe the results following Moser [2]. If A is a finite or countably infinite sequence of symbols, let S be the collection of doubly infinite sequences $s = \{s_k, k = 0, \pm 1, \ldots\}$ with each $s_k \in A$. The shift automorphism σ on S is defined by $\sigma s = \{\tilde{s}_k, k = 0, \pm 1, \ldots\}$, $\tilde{s}_k = s_{k-1}$ for all k.

Near a homoclinic point q, we can construct a small quadrilateral Q, two of its sides consisting of parts of $W_r^u(p)$, $W_r^s(p)$ and the others parallel to the tangents of these sets at q (see Figure 1). For any point $\alpha \in Q$, let $k = k(\alpha)$ be the smallest positive integer such that

$\pi^k(\alpha) \in Q$, if it exists. Let $D(\widetilde{\pi})$ be the set of $\alpha \in Q$ for which such a k exists and define $\widetilde{\pi}\alpha = \pi^k(\alpha)$ for $\alpha \in D(\widetilde{\pi})$. The map $\widetilde{\pi}$ is called the *transversal map* of π for the quadrilateral Q.

THEOREM 3.8. *If π is a C^∞-diffeomorphism of the plane with a point q transverse homoclinic to a hyperbolic fixed point p, then in a neighborhood of q, the transversal map $\widetilde{\pi}$ of a quadrilateral possesses an invariant set I homeomorphic to the sequence space S with an infinite number of symbols by a map $\tau: S \longrightarrow I$ such that $\widetilde{\pi}\tau = \tau\sigma$. Also, there is an integer k, an invariant set $\widetilde{I}$ of π^k and a homeomorphism $\widetilde{\tau}: S \longrightarrow \widetilde{I}$, where S is the sequence space of a finite number of symbols, such that $\pi^k\widetilde{\tau} = \widetilde{\tau}\sigma$.*

Note the difference in the two conclusions in the theorem. In the first part, the set I is invariant for $\widetilde{\pi}$ and $\widetilde{\pi}$ is equivalent on I to the shift automorphism on an infinite number of symbols. In the second part, the set $\widetilde{I}$ is invariant under a fixed power k of π itself and π^k is equivalent on $\widetilde{I}$ to the shift automorphism on a finite number of symbols.

It follows immediately from Theorem 3.8 that there are infinitely many periodic points in a neighborhood of the transverse homoclinic point and they are dense. Also, there is a random behavior to the orbits on the invariant set I (or $\widetilde{I}$) since knowing the early terms of a sequence tells nothing about the later terms of a sequence.

For examples of transverse homoclinic points in celestial mechanics, see Moser [2]. Transverse homoclinic points also occur in structurally stable systems as we shall see below. More examples in second order nonautonomous differential equations will be given later when we are studying analytical methods in bifurcation theory. Now, we prefer to continue the general survey.

To describe further aspects of the theory, it is convenient to work with $\mathrm{Diff}^r(M)$, $r \geqslant 1$, the space of diffeomorphisms with derivatives up through order r on a smooth compact manifold M. This can be related to differential equations in several ways. One of the most important is through a Poincaré map for periodic orbits as described above. More generally, if $f \in \mathcal{X}^r_n(\bar{\Omega})$, $M \subset \bar{\Omega}$ is compact and, for each $x \in M$, there is a $\tau(x) > 0$ such that $T_f(\tau(x))x \in M$, $T_f(t)x \notin M$, $0 < t < \tau(x)$, then the map $x \mapsto T_f(\tau(x))x$ is in $\mathrm{Diff}^r(M)$ if it has the required number of derivatives.

If $g \in \mathrm{Diff}^r(M)$, a point $p \in M$ is a periodic point of g if there is a positive integer $n = n(p)$ such that $T^n p = p$. The periodic orbit is *hyperbolic* if no eigenvalue of $\partial g(p)/\partial x$ has modulus one. For each hyperbolic periodic point p of g, one can define the global stable manifold $W^s(p)$ and unstable manifold $W^u(p)$ in a manner similar to the definitions for vector fields.

We now give an example due to Thom which was an inspiration for many further developments in dynamical systems. In $\mathbf{R}^2$ identify the points (x, y), $(x + m, y + n)$ for all integers m, n. Any unit square with integer vertices may be identified with the torus T^2 and any mapping of the plane into itself yields a mapping of T^2 into T^2 in the obvious way. Let L be a 2×2 matrix with integer coefficients, determinant 1 and real eigenvalues. The eigenvalues are then λ, λ^{-1} with $\lambda < 1$ irrational. This implies that the linear subspaces E^s, E^u generated respectively by the eigenvectors for λ, λ^{-1} have irrational slope. For any

$x \in \mathbf{R}^2$, each of the lines $x + E^s$, $x + E^u$ is invariant under L. The map L on $\mathbf{R}^2$ generates a natural map π on T^2 obtained from the above identification of T^2 with unit squares in $\mathbf{R}^2$ with integer coefficients. If $p = \pi x \in T^2$, $x \in \mathbf{R}^2$, let $W^s(p) = \pi(x + E^s)$, $W^u(p) = \pi(x + E^u)$. If p is a periodic point of π, then $W^s(p)$, $W^u(p)$ are respectively the stable and unstable manifolds of p. Since the slopes of the linear subspaces E^s, E^u are irrational, the sets $W^s(p)$ and $W^u(p)$ are dense in T^2 for every $p \in T^2$. Furthermore, it is not difficult to show that every point of intersection of these sets is a point of transversal intersection. Also, the points of intersection are dense in T^2. In particular, there is a dense set of points transverse homoclinic to the critical point $p = \pi(0)$. Since each transverse homoclinic point is the limit of periodic points, it follows that the periodic points of π are dense in T^2. A simple direct proof of this last result is contained in Palis and deMelo [1, p. 171].

With $\pi\colon T^2 \longrightarrow T^2$ defined as above, Anosov [1], [2] showed that π is structurally stable. A more elementary proof was given by Moser [1], [2]. Since π is structurally stable and contains infinitely many periodic orbits, this necessarily implies that the Morse-Smale systems are not dense in the set of structurally stable systems.

The above example was generalized by Anosov [2] in the following way.

DEFINITION 3.9. Let M be a compact manifold. An $f \in \mathrm{Diff}^r(M)$, $r \geqslant 1$, is an *Anosov diffeomorphism* if the tangent space at each point x of M is a direct sum $E_x^s \oplus E_x^u$ invariant under the derivative Df; that is, $Df_x E_x^s = E_{f(x)}^s$, $Df_x E_x^u = E_{f(x)}^u$ and there is a Riemannian metric on M and a constant $\lambda \in (0, 1)$ such that $|Df_x v| \leqslant \lambda |v|$, $|Df_x^{-1} u| \leqslant \lambda |u|$ for all $x \in M$, $v \in E_x^s$, $u \in E_x^u$.

Anosov [2] has shown that these diffeomorphisms are structurally stable. A simpler proof was given by Moser [1], [2]. For a discussion of the restrictions that are imposed on the manifold M in order for it to admit an Anosov diffeomorphism, see Palis and deMelo [1].

The next important step in the abstract theory of dynamical systems was taken by Smale [2] by defining systems which satisfy Axiom A. Suppose $f \in \mathrm{Diff}^r(M)$ and $\Lambda \subset M$ is a closed invariant set. The set Λ is said to have a hyperbolic structure if the tangent space at each point $x \in \Lambda$ is the direct sum $E_x^s \oplus E_x^u$ invariant under Df and there are a Riemannian metric and $\lambda \in (0, 1)$ such that $|Df_x v| \leqslant \lambda |v|$, $|Df_x^{-1} u| \leqslant \lambda |u|$ for $x \in \Lambda$, $v \in E_x^s$, $u \in E_x^u$. If Λ is hyperbolic, it is possible to define stable and unstable manifolds for the set Λ by looking at asymptotic orbits.

DEFINITION 3.10. $f \in \mathrm{Diff}^r(M)$ satisfies *Axiom A* if the set of nonwandering points $\Omega(f)$ is hyperbolic and the periodic points of f are dense in $\Omega(f)$.

If f satisfies Axiom A, Smale [2] has shown that $\Omega(f) = \Omega_1 \cup \Omega_2 \cup \cdots \cup \Omega_k$ where each Ω_j is closed invariant and transitive; that is, has a dense orbit. Robinson [1] has shown that any $f \in \mathrm{Diff}^r(M)$, $r \geqslant 1$, is structurally stable if it satisfies Axiom A and all stable and unstable manifolds intersect transversally. An $f \in \mathrm{Diff}^r(M)$ is said to be *absolutely stable* if there are a neighborhood $V(f) \subseteq \mathrm{Diff}^r(M)$ of f and a constant $K > 0$ such that, for every $g \in V(f)$, there is a homeomorphism h of M such that $hf = gh$ and $|h - I|_0 < K|f - g|_0$ where $|\cdot|_0$ designates the norm in C^0. Results of Franks [1], Guckenheimer [1] and Mañé [1] show that f is absolutely stable if and only if it satisfies Axiom A and all stable and unstable manifolds intersect transversally. Mañé (unpublished) has also recently shown that Axiom A is implied by structural stability and a technical condition on the characteristic exponents of Liapunov on $\Omega(f)$.

It is also possible to study structural stability restricted to the set of nonwandering points. More specifically, $f \in \mathrm{Diff}^r(M)$ is said to be *Ω-stable* if there is a neighborhood $V(f)$ of f such that, for every $g \in V(f)$, there is a homeomorphism $h: \Omega(f) \longrightarrow \Omega(g)$ such that $hf = gh$ on $\Omega(f)$. If f satisfies Axiom A, $\Omega(f) = \Omega_1 \cup \cdots \cup \Omega_k$, then a *cycle* of Ω is a sequence $p_1 \in \Omega_{k_1}, \ldots, p_s \in \Omega_{k_s} = \Omega_{k_1}$ such that $W^s(p_i) \cap W^u(p_{i+1}) \neq \emptyset$, $1 \leqslant i \leqslant s-1$. Smale [2] showed that Axiom A and no cycles imply Ω-stable. Palis [2] has shown that any f satisfying Axiom A is not Ω-stable if it has a cycle. It is not known if f Ω-stable implies that it must satisfy Axiom A.

4. Stability and bifurcation at a zero eigenvalue

Suppose A is an $n \times n$ constant matrix whose eigenvalues have negative real parts, $x \in \mathbf{R}$, $y \in \mathbf{R}^n$, $X_0 \in C^k(\mathbf{R} \times \mathbf{R}^n, \mathbf{R})$, $Y_0 \in C^k(\mathbf{R} \times \mathbf{R}^n, \mathbf{R}^n)$, $k \geqslant 1$, are functions of $(x, y) \in \mathbf{R} \times \mathbf{R}^n$ which vanish together with their first derivatives at $x = 0, y = 0$. One of our objectives in this section is to determine how the stability properties of the solution $x = 0, y = 0$, of the equation

$$\dot{x} = X_0(x, y), \qquad \dot{y} = Ay + Y_0(x, y) \tag{4.1}$$

depend upon the nonlinear functions X_0, Y_0.

For a given neighborhood U of $x = 0, y = 0$, and $X \in C^k(U, \mathbf{R})$, $Y \in C^k(U, \mathbf{R})$, $k \geqslant 1$, we also discuss how the number of equilibrium solutions of the perturbed equation

$$\dot{x} = X_0(x, y) + X(x, y), \qquad \dot{y} = Ay + Y_0(x, y) + Y(x, y) \tag{4.2}$$

depends upon X, Y in a neighborhood of $X = 0$, $Y = 0$. The stability properties of these solutions will be considered. A complete classification is given for some cases. The neighborhood U will be fixed and our analysis always will be in some other neighborhood V of $x = 0, y = 0$ which will be a subset of U.

LEMMA 4.1. *There are a neighborhood V of $(x, y) = (0, 0)$ and a neighborhood W of $(X, Y) = (0, 0)$ such that, for $(X, Y) \in W$ and any equilibrium point $\alpha = \alpha(X, Y)$ of* (4.2) *in V, the stable manifold has dimension $\geqslant n$. Furthermore, there is an n-dimensional manifold $S(\alpha, X, Y)$ in V containing α such that any solution of* (4.2) *with initial value in $S(\alpha, X, Y)$ approaches α as $t \longrightarrow \infty$ exponentially. Furthermore, $S(\alpha, X, Y)$ is C^k in (X, Y) and the tangent space to $S(\alpha, X, Y)$ at α approaches the space $\{(0, y), y \in \mathbf{R}^n\}$ as $(X, Y) \longrightarrow (0, 0)$.*

A proof will not be given. It is technical but may be supplied by using the ideas in Coddington and Levinson [1], Hale [1] or Hartman [2] for the construction of stable manifolds.

For equation (4.2), a set $S \subset \mathbf{R}^{n+1}$ is a *local invariant manifold* if, for any $(x_0, y_0) \in S$, there is a $T > 0$ such that the solution of (4.2) through (x_0, y_0) remains in S for $|t| < T$. If $T = \infty$, we say S is an *invariant manifold.* If there is a C^1 function $\psi(x, X, Y)$ defined in a neighborhood of $(x, X, Y) = (0, 0, 0)$ such that $\psi(0, 0, 0) = 0$, $\partial\psi(0,0,0)/\partial x = 0$ and $S = S(X, Y) = \{(x, y): y = \psi(x, X, Y)\}$ is a local invariant manifold for (4.2), then we say S is a *local center manifold for* (4.2).

Because the matrix A has eigenvalues with negative real parts, it is reasonable to expect that there is a local center manifold for (4.2). This idea is implicit in the papers of Krylov and Bogoliubov written in the 1930's (see Hale [1] for references) and was used explicitly by V. Pliss [1] to study stability. Kelley [1], [2] has given a very complete and readable account of this idea. More precisely, he has shown the existence of a center manifold which is C^k in (x, X, Y) in a sufficiently small neighborhood V (depending on k) of $(x, y) = (0, 0)$. A center manifold is also exponentially stable in the sense that any solution which remains in the neighborhood V of $(x, y) = (0, 0)$ approaches the center manifold exponentially as $t \longrightarrow \infty$.

The flow on a center manifold $S(X, Y)$ is given by the scalar equation

$$\begin{aligned} \dot{x} &= u(x, X, Y), \\ u(x, X, Y) &= X_0(x, \psi(x, X, Y)) + X(x, \psi(x, X, Y)). \end{aligned} \tag{4.3}$$

If α is an equilibrium point of equation (4.2), then $\alpha \in S(X, Y)$, $\alpha = (x_0, \psi(x_0, X, Y))$ for some x_0. The converse is also true. The stability properties of x_0 as a solution of (4.3) are easily determined from the behavior of $u(x, X, Y)$ near $x = x_0$. From Lemma 4.1, the stability properties of x_0 are reflected in an obvious way in the stability properties of α as a solution of (4.2). In particular, x_0 being a stable (asymptotically stable) (unstable) equilibrium point of (4.3) is equivalent to α being a stable (asymptotically stable) (unstable) equilibrium point of (4.2).

From the above discussion, the complete structure of the flow for (4.2) is determined by knowing the flow for (4.3). Thus, if we can determine a center manifold, the problem will be solved. Generally, this is extremely difficult to do. We give an alternative procedure for determining the qualitative properties of the solutions of (4.2) which is very close to the original method of Liapunov [1]. It uses only the bifurcation function obtained by an application of the method of Liapunov-Schmidt for the equilibrium solutions of equation (4.2).

This method has other advantages over the use of the center manifold. The bifurcation function has the same smoothness properties as the original vector field. This is not true of the local center manifold. It inherits the C^k property for any finite k, but may not be C^∞ (or analytic even) if the original vector fields are C^∞ (or analytic). For examples, see Carr [1].

There is a neighborhood of $(0, 0) \in \mathbf{R} \times \mathbf{R}^n$ and a neighborhood of $Y = 0$ such that the equation

$$A\phi + Y_0(x, \phi) + Y(x, \phi) = 0 \tag{4.4}$$

has a unique solution $\phi(x, Y)$ with $\phi(0, 0) = 0$. This function is C^k in (x, Y). Let

$$G(x, X, Y) = X_0(x, \phi(x, Y)) + X(x, \phi(x, Y)). \tag{4.5}$$

We can now prove the following result.

THEOREM 4.2. *There are neighborhoods* V_1 *of* $x = 0$, V_2 *of* $y = 0$, W_1 *of* $X = 0$, W_2 *of* $Y = 0$ *and a* C^k*-function* ϕ: $V_1 \times W_2 \longrightarrow V_2$ *such that the equilibrium points* (x_0, y_0) *of* (4.2) *in* $V_1 \times V_2$ *for* $(X, Y) \in W_1 \times W_2$ *are given by*

$$G(x_0, X, Y) = 0, \qquad y_0 = \phi(x_0, Y) \tag{4.6}$$

where G is defined in (4.5). *Furthermore, the stability properties of the solution* x_0 *of the scalar equation*

$$\dot{x} = G(x, X, Y) \tag{4.7}$$

are the same as the stability properties of x_0 *on a center manifold. Finally, any solution of* (4.2) *which remains in the neighborhood* $V_1 \times V_2$ *for* $t \geqslant 0$ *approaches an equilibrium point of* (4.2).

The function $G(x, X, Y)$ is easier to compute than the vector field $u(x, X, Y)$ on a center manifold. The importance of the result is that the function $G(x, X, Y)$, called the *bifurcation function,* carries all of the information necessary to determine the stability properties of the equilibrium points on the center manifold. The conclusion on stability in Theorem 4.2 is due to deOliveira and Hale [1].

PROOF OF THEOREM 4.2. It is clear that the equilibrium points of (4.2) are given by (4.6). If $u(x, X, Y)$ is the vector field on a center manifold, then $u(x, X, Y)$ and $G(x, X, Y)$ have the same set of zeros in a neighborhood of zero. If x_0 is an isolated zero, let us first prove that G, u have the same sign in a neighborhood of x_0. Without loss in generality, we can suppose $x_0 = 0$ and it is simple because we can make a small perturbation in X, Y to attain this property. So, assume $G(0, X, Y) = 0 = u(0, X, Y)$, $\partial G(0, X, Y)/\partial x > 0$, $\partial u(0, X, Y)/\partial x < 0$. Now, consider the functions $G(x, X + \epsilon, Y)$, $u(x, X + \epsilon, Y)$ for ϵ a small real number. Then one can show that

$$G(x, X + \epsilon, Y) = G(x, X, Y) + \epsilon,$$

$$u(x, X + \epsilon, Y) = u(x, X, Y) + \delta(x, X, Y)\epsilon + o(|\epsilon|)$$

as $|\epsilon| \longrightarrow 0$ where $\delta(x, X, Y) > 0$ for x in a neighborhood of zero. The assertion about G is obvious, but the one about u is nontrivial and uses the properties of a center manifold. Since the derivatives of G and u with respect to x at $(0, 0, 0)$ have opposite sign, this shows that a small perturbation of ϵ gives rise to distinct zeros of G, u in a neighborhood of $x = 0$. However, this is a contradiction since G, u must always have the same set of zeros. Thus, if x_0 is an isolated zero of G, u, then the functions G, u have the same sign in a neighborhood of zero. Thus, the stability properties as a solution of (4.7) and (4.3) are the same. If x_0 is not isolated, then one can easily discuss all the ways in which this can occur to complete the discussion of the stability of x_0.

If a positive orbit stays in $V_1 \times V_2$, then its ω-limit set must be on the center manifold and it must be a connected invariant set. Also, Lemma 4.1 implies that each equilibrium point has a stable manifold of codimension one transverse to the center manifold. If the ω-limit set contains two distinct points, then it must contain the arc between these two points on the center manifold. If there are no equilibrium points between these two, then the positive orbit must remain between the transverse stable manifolds above and go back and forth along the arc as $t \longrightarrow \infty$, an obvious contradiction. The same type of argument can be used even if there are any number of points in the arc. The theorem is proved.

The proof of the last statement in the theorem can be modified following Hale and Massatt [1] to obtain the following result.

THEOREM 4.3. *The ω-limit set of a bounded orbit of the gradient system $\dot{x} = f(x)$, $f(x) = \text{grad } F(x)$ is a single point if, for every zero x_0 of f which belongs to a continuum of zeros of f, the matrix $\partial f(x_0)/\partial x$ is either nonsingular or has zero as a simple eigenvalue.*

PROOF. We only indicate a proof. Details may be found in Hale and Massatt [1]. Differentiating $F(x)$ along the solutions one sees that the ω-limit set of every solution must belong to the set of equilibrium points. Since the ω-limit set is connected, it must contain either a single equilibrium point or a continuum of such points. If x_0 belongs to a continuum of the ω-limit set, then the hypothesis on $\partial f(x_0)/\partial x$ implies the existence of a smooth center manifold through x_0 of dimension one. Thus, the ω-limit set near x_0 is a smooth arc. Since it is compact, it must be an arc or a closed curve. But then one can use a generalization of Lemma 4.1 to show there is a tubular neighborhood of the ω-limit set such that any orbit which remains in this neighborhood for all $t \geqslant 0$ lies on a stable manifold for an equilibrium point. This proves the result.

As a consequence of Theorem 4.2, we have

COROLLARY 4.4. *An equilibrium solution $(x_0, y_0) \in V_1 \times V_2$ of (4.2) for $(X, Y) \in W_1 \times W_2$ is asymptotically stable if and only if there is an $\epsilon > 0$ such that $G(x_0 + u, X, Y)u < 0$ for $0 < |u| < \epsilon$. It is unstable if and only if there is an $\epsilon > 0$ such that $G(x_0 + u, X, Y)u > 0$ for either $0 < u < \epsilon$ or $-\epsilon < u < 0$. If there is an $\epsilon > 0$ such that $G(x_0 + u, X, Y) = 0$ for $0 \leqslant |u| < \epsilon$, then the solution is stable and there is a C^k first integral in a neighborhood of (x_0, y_0). If the functions X_0, X, Y_0, Y are analytic (or C^∞) then this first integral is analytic (or C^∞).*

PROOF. Everything is obvious except the existence of the first integral. Lemma 4.1 implies the existence of an n-dimensional C^k-stable manifold above each equilibrium point $x_0 + u$, $|u| < \epsilon$. These manifolds are analytic if the functions in (4.2) are analytic. Using these stable manifolds, one can make a C^k (or analytic) change of variables $(x, y) \mapsto (v, w)$ so that equation (4.2) in a neighborhood of (x_0, y_0) becomes $\dot{v} = 0$, $\dot{w} = Aw + W(v, w)$. This equation has a first integral $V(v, w) = v$. Thus, the original equation has a first integral and the proof is complete.

COROLLARY 4.5. *Suppose $G(x_0, X, Y) = 0$ and $y_0 = \phi(x_0, Y)$, ϕ satisfying (4.4). If there is an integer $q \geqslant 1$ and a $\beta \neq 0$ such that*

$$G(x_0 + u, X, Y) = \beta u^q + o(|u|^q) \quad \text{as } |u| \longrightarrow 0, \tag{4.8}$$

then the solution (x_0, y_0) *of* (4.2) *is asymptotically stable if and only if* $\beta < 0$, q *odd. Otherwise, it is unstable.*

We have also the following results of Liapunov [1] (see, also, Bibikov [1]).

COROLLARY 4.6. *Suppose* $G(x_0, X, Y) = 0$, $y_0 = \phi(x_0, Y)$, ϕ *satisfying* (4.4) *and* X_0, X, Y_0, Y *are analytic. Then either there are an integer* $q \geq 1$ *and a* $\beta \neq 0$ *such that* (4.8) *is satisfied with the stability properties as stated in Corollary* 4.5 *or* (x_0, y_0) *is a stable solution of* (4.2) *and there is an analytic first integral.*

PROOF. Since $G(x, X, Y)$ is analytic, only the two cases mentioned can occur.

COROLLARY 4.7. *If there is a scalar function* $H(x, y, z)$ *continuous for* $(x, y, z) \in \mathbf{R} \times \mathbf{R}^n \times \mathbf{R}^n$ *such that* $H(x, y, 0) = 0$ *and*

$$X_0(x, y) + X(x, y) = H(x, y, Ay + Y_0(x, y) + Y(x, y))$$

then the zero solution of (4.2) *is stable and there is a first integral.*

PROOF. The hypothesis implies $G(x, X, Y) = 0$ for x in a neighborhood of $X = 0$.

Let us now turn to the problem of bifurcation of the equilibrium point (0, 0) treating X, Y in (4.2) as parameters. An equilibrium point (x_0, y_0) of (4.2) is called a *saddle-node* if x_0 as a solution of (4.7) or (4.3) is asymptotically stable from one side and unstable from the other. The following result is contained in Andronov et al. [1], Sotomayor [1].

THEOREM 4.8. *Suppose* $k \geq 2$ *and there is a* $\beta \neq 0$ *such that*

$$G(x, 0, 0) = \beta x^2 + o(|x|^2) \quad as\ |x| \longrightarrow 0. \tag{4.9}$$

Then there is a neighborhood V *of* $(x, y) = (0, 0)$, *a neighborhood* W *of* $(X, Y) = 0$ *and a* C^k*-function* γ: $W \longrightarrow \mathbf{R}$ *such that the following conclusions hold*:

(i) $\gamma(X, Y) > 0$ *implies there are no equilibrium points of* (4.2) *in* V.

(ii) $\gamma(X, Y) = 0$ *implies there is one equilibrium point of* (4.2) *in* V *which is a saddle-node.*

(iii) $\gamma(X, Y) < 0$ *implies there are two equilibrium points of* (4.2) *in* V, *one is a hyperbolic saddle and the other is hyperbolic and asymptotically stable.*

Furthermore, the set $\Gamma = \{(X, Y): \gamma(X, Y) = 0\}$ *is a submanifold of codimension* 1 *in* W *(see Figure* 2).

PROOF. The function $G(x, X, Y)$ satisfies $\partial G(0, 0, 0)/\partial x = 0$, $\partial^2 G(0, 0, 0)/\partial x^2 = 2\beta \neq 0$. Thus, there are a neighborhood W of $(X, Y) = (0, 0)$ and a neighborhood V_1 of $x = 0$ such that the equation $\partial G(x, X, Y)/\partial x = 0$ has a unique solution $x^*(X, Y)$ with $x^*(0, 0) = 0$. If $\gamma(X, Y) = (\operatorname{sgn} \beta)G(x^*(X, Y), X, Y)$, then the conclusion of the first part of the theorem follows from Theorem 4.2.

To prove the last part of the theorem, consider the special case where $X(x, y) = \lambda$, $Y(x, y) = 0$ for all x, y where λ is a real scalar and show that $D_\lambda \gamma(\lambda, 0) \neq 0$ for $\lambda = 0$.

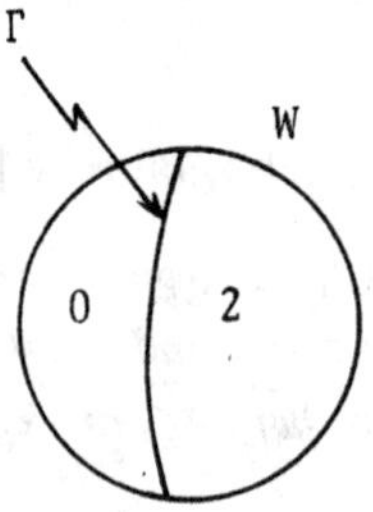

FIGURE 2.

$G(x, 0, 0) = \beta x^2 + o(|x|^2)$.

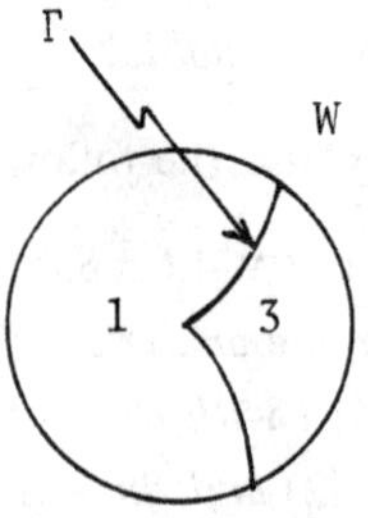

FIGURE 3.

$G(x, 0, 0) = \beta x^3 + o(|x|^3)$.

To discuss a higher order bifurcation, suppose that $k \geqslant 3$ and

$$G(x, 0, 0) = \beta x^3 + o(|x|^3) \quad \text{as } |x| \to 0, \ \beta \neq 0. \tag{4.10}$$

The argument below follows Chow, Hale and Mallet-Paret [1] and Vanderbauwhede [1]. Then there are a neighborhood V_1 of $x = 0$ and a neighborhood W of $(X, Y) = (0, 0)$ such that the equation $\partial^2 G(x, X, Y)/\partial x^2 = 0$ has a unique solution $x_1^*(X, Y)$ with $x_1^*(0, 0) = 0$. Let

$$\gamma_1(X, Y) = \frac{\partial G(x_1^*(X, Y), X, Y)}{\partial x}. \tag{4.11}$$

If we apply the same argument as in the proof of Theorem 4.8, then for $(X, Y) \in W$, the equation $\partial G(x, X, Y)/\partial x$ has no solution in V_1 if $\beta\gamma_1(X, Y) > 0$, one solution if $\gamma_1(X, Y) = 0$ and two simple solutions if $\beta\gamma_1(X, Y) < 0$.

If $\beta\gamma_1(X, Y) > 0$, then $G(x, X, Y)$ is strictly increasing and has exactly one solution in V_1. If $\gamma_1(X, Y) = 0$, let $x_2^*(X, Y)$ be the unique solution of $\partial G(x, X, Y)/\partial x = 0$ in V_1 and define

$$\gamma_2(X, Y) = G(x_2^*(X, Y), X, Y). \tag{4.12}$$

If $\gamma_1(X, Y) = 0$, $\gamma_2(X, Y) = 0$, then the unique solution of $G(x, X, Y) = 0$ in V_1 is a triple zero. If either $\beta\gamma_1(X, Y) > 0$ or $\gamma_1(X, Y) = 0$, $\gamma_2(X, Y) \neq 0$, then this unique solution is simple.

Now suppose $\beta\gamma_1(X, Y) < 0$. Then there are two simple solutions of $\partial G(x, X, Y)/\partial x = 0$ in V_1. Again using the argument of the proof of Theorem 4.8, one can show that they have the form

$$x_\pm^* = x_2^*(X, Y) + \sigma_\pm^*(X, Y, \gamma_1(X, Y))(-\beta\gamma_1(X, Y))^{1/2}$$

where $\sigma_\pm^*(X, Y, 0) = \pm 3^{-1/2}$ (see Vanderbauwhede [1] for details).

If

$$\gamma(X, Y) = G(x_+^*(X, Y), X, Y)G(x_-^*(X, Y), X, Y) \tag{4.13}$$

then, using Theorem 4.2, we can state the following theorem.

THEOREM 4.9. *Suppose* $G(x, 0, 0)$ *satisfies* (4.10), *the regions* V_1, W *and functions* γ_1, γ *on* W *are defined as above. Then the following conclusions hold*:

(i) *If either* $\beta\gamma_1(X, Y) \geqslant 0$ *or* $\beta\gamma_1(X, Y) < 0$ *and* $\gamma(X, Y) > 0$, *then equation* (4.6) *has one zero in* V *and equation* (4.2) *has exactly one equilibrium point in a neighborhood of zero which is asymptotically stable if* $\beta < 0$ *and unstable if* $\beta > 0$.

(ii) *If* $\beta\gamma_1(X, Y) < 0$ *and* $\gamma(X, Y) = 0$, *then equation* (4.6) *has two zeros and equation* (4.2) *has two equilibrium points in a neighborhood of zero, one being a saddle-node and the other hyperbolic. The hyperbolic one is asymptotically stable if* $\beta < 0$ *and a saddle point of order one if* $\beta > 0$.

(iii) *If* $\beta\gamma_1(X, Y) < 0$ *and* $\gamma(X, Y) < 0$, *there are three simple solutions of equation* (4.6) *and equation* (4.2) *has three equilibrium points in a neighborhood of zero, all hyperbolic, two saddles of order one if* $\beta > 0$ *and one saddle of order one if* $\beta < 0$, *the others being asymptotically stable.*

In the above theorem, the set $\Gamma = \{(X, Y): \beta\gamma_1(X, Y) < 0, \gamma(X, Y) = 0\}$ is the bifurcation set in W; that is, the set where the number of equilibrium points of (4.2) changes from one to three and where the topological structure of the trajectories of (4.2) changes in a neighborhood of zero.

It is possible to show that there are positive nonzero functions σ_+, σ_- on W such that saying that $(X, Y) \in \Gamma$ implies either

$$\gamma_2(X, Y) = \sigma_+(X, Y)(-\beta\gamma_1(X, Y))^{3/2}$$

or

$$\gamma_2(X, Y) = -\sigma_-(X, Y)(-\beta\gamma_1(X, Y))^{3/2}$$

where γ_2 is defined in (4.12). This shows that the set Γ is like a cusp surface in W as in Figure 3 (see Vanderbauwhede [1] for details).

With G satisfying (4.9), Theorem 4.8 implies the complete behavior of the solutions of (4.2) in a neighborhood of zero is characterized by one function of X, Y. If G satisfies (4.10), then two functions were needed to obtain this characterization. If the degeneracy of $G(x, X, Y)$ at $(X, Y) = (0, 0)$ is higher order, then it becomes extremely difficult to give an explicit description of how the structure of the zeros depends on X, Y. This is due to the fact that it is always difficult to discuss the zeros of polynomials of degree greater than three.

Singularity theory can be used to say that the problem actually is reducible to the discussion of a polynomial. If we suppose all functions in (4.2) are C^∞, then $G(x, X, Y)$ is C^∞ in x, X, Y. If

$$G(x, 0, 0) = \beta x^q + o(|x|^{q+1}) \quad \text{as } |x| \to 0, \ \beta \neq 0, \tag{4.14}$$

then the Banach space version of the Malgrange Preparation Theorem due to Michor [1] implies there are a polynomial $P(x, X, Y)$, $P(x, 0, 0) = \beta x^q$ of degree q in x, and a positive function $E(x, X, Y)$, $E(x, 0, 0) = 1$, such that

$$G(x, X, Y) = E(x, X, Y)P(x, X, Y) \tag{4.15}$$

and these functions are C^∞ in x, X, Y. Thus the zeros of $G(x, X, Y)$ as well as the signs between zeros coincide with those of $P(x, X, Y)$. Thus, we can state an analogue of Theorem 4.2 in the following way. The equilibrium points of (4.2) are given by

$$P(x_0, X, Y) = 0, \quad y = \phi(x_0, Y) \tag{4.16}$$

and the stability properties of the equilibrium points are determined by the stability properties of the equilibrium point x_0 of the scalar equation

$$\dot{x} = P(x, X, Y). \tag{4.17}$$

Such a result could not be obtained by reducing the discussion of (4.2) to the center manifold (4.3) and then applying the Malgrange theorem. The reason is that the center manifold is not necessarily C^∞ (see Carr [1]).

For the special case $q = 2$, one can take $P(x, X, Y) = \beta x^2 + \gamma(X, Y)$ where γ is the function in Theorem 4.9. For $q = 3$, one can take $P(x, X, Y) = \beta x^3 + \gamma_1(X, Y)x + \gamma_3(X, Y)$ where γ_3 is essentially the same function as γ_2 in (4.12).

In the applications, one frequently encounters the situation where the perturbed vector field (X, Y) is not an arbitrary function in a neighborhood of zero, but belongs to a prescribed k parameter family of vector fields $X(x, y, \lambda)$, $Y(x, y, \lambda)$, $\lambda \in \mathbf{R}^k$, which vanish for $\lambda = 0$. The bifurcation function is then a function of λ, $G = G(x, \lambda)$. The number of equilibrium points and their stability properties are determined by the manner in which the k parameter family of vector fields crosses the bifurcation surfaces obtained from taking arbitrary perturbations X, Y. Of course, it is clear that the same conclusions will prevail if the original problem is phrased in terms of the restricted family depending on λ. This latter method is often more transparent.

As a final remark, it is possible to extend everything above to the case where the eigenvalues of A in equation (4.1) have nonzero real parts.

5. Stability and bifurcation from a focus

In this section, we discuss bifurcation from an equilibrium point in the spirit of the previous section except under the hypothesis that the linear approximation of the vector field has two purely imaginary roots and the remaining ones have negative real parts.

The unperturbed equation is given as

$$(5.1)\qquad \dot{x} = Bx + X_0(x, y), \qquad \dot{y} = Ay + Y_0(x, y)$$

where

$$B = \begin{bmatrix} 0 & 1 \\ -1 & 0 \end{bmatrix}.$$

A is an $n \times n$ matrix whose eigenvalues have negative real parts, $X_0 \in C^k(\mathbf{R}^2 \times \mathbf{R}^n, \mathbf{R}^2)$, $Y_0 \in C^k(\mathbf{R}^2 \times \mathbf{R}^n, \mathbf{R}^n)$, $k \geqslant 1$, are functions vanishing together with their first derivatives at $x = 0, y = 0$.

For a given neighborhood U of $x = 0, y = 0$ and $X \in C^k(U, \mathbf{R}^2)$, $Y \in C^k(U, \mathbf{R}^n)$, the perturbed equation is

$$(5.2)\qquad \dot{x} = Bx + X_0(x, y) + X(x, y), \qquad \dot{y} = Ay + Y_0(x, y) + Y(x, y).$$

For (X, Y) in a sufficiently small neighborhood of $(0, 0)$, equation (5.2) has a unique equilibrium point $x = \alpha(X, Y), y = \beta(X, Y)$ in a neighborhood of $x = 0, y = 0, \alpha(0, 0) = 0$, $\beta(0, 0) = 0$. The functions α, β are C^k-functions of X, Y. By a translation of variables, we can therefore assume that this equilibrium point is zero. Thus, we will assume without loss of generality that

$$(5.3)\qquad X(0, 0) = 0, \qquad Y(0, 0) = 0.$$

Our objective is to study how the stability properties of the unperturbed equation (5.1) are determined by the nonlinearities X_0, Y_0. Also, we determine the number and stability properties of the periodic orbits that can bifurcate from zero as the perturbation terms X, Y are varied in a neighborhood of zero. The existence and the number of periodic solutions follow from the bifurcation function obtained by an application of the method of Liapunov-Schmidt. We prove that this bifurcation function also carries the information on the stability of periodic orbits. The proof of this latter fact uses the center manifold theorem.

LEMMA 5.1. *There are a neighborhood V of $(x, y) = (0, 0)$ and a neighborhood W of $(X, Y) = (0, 0)$ such that, for $(X, Y) \in W$ and any periodic orbit $\gamma = \gamma(X, Y)$ of (5.2) in V, the stable manifold has dimension $\geqslant n + 1$. Furthermore, there is an $(n + 1)$-dimensional manifold $S(\gamma, X, Y)$ (either a generalized Möbius band or the cross-product of a circle and an n-dimensional ball) containing γ such that γ is exponentially asymptotically orbitally stable with asymptotic phase relative to initial values in $S(\gamma, X, Y)$. Furthermore, $S(\gamma, X, Y)$ is C^k in X, Y and is diffeomorphic to the local stable manifold for any periodic orbit of the linear equation $\dot{x} = Bx$, $\dot{y} = Ay$.*

The proof of this result may be supplied by using the ideas for the construction of stable manifolds near periodic orbits in Coddington and Levinson [1], Hale [1] (see, also, Fenichel [1]).

The manifold $S(\gamma, X, Y)$ in Lemma 5.1 has codimension one. Thus, the complete behavior of the solutions near a periodic orbit is determined by what happens in one other direction. An analysis of the Poincaré map on a center manifold will take care of this direction.

There is a center manifold for equaiton (5.2) which is C^k in (x, X, Y) in a sufficiently small neighborhood V of $(x, y) = (0, 0)$ and a neighborhood W of $(X, Y) = (0, 0)$. If a center manifold is given by $S = \{(x, y): y = \psi(x, X, Y)\}$ then the flow on the center manifold S is given by

$$\begin{aligned} &\dot{x} = Bx + \widetilde{X}(x), \\ (5.4)\qquad &\widetilde{X}(x) = X_0(x, \psi(x, X, Y)) + X(x, \psi(x, X, Y)). \end{aligned}$$

If γ is a periodic orbit of (5.2), then γ belongs to S and conversely. Thus, the periodic orbits of (5.2) can be determined by discussing the periodic orbits of (5.4). Also, from Lemma 5.1, the stability properties of a periodic orbit of (5.2) are completely determined by the stability properties of the corresponding periodic orbit of (5.4).

By the manner in which a center manifold is constructed, one can also obtain a priori bounds on ψ. In fact, for any $\epsilon > 0$, $\delta > 0$, there is a constant $K = K(\epsilon, \delta)$ such that, if $y = \psi(x, X, Y)$ is a center manifold of (5.2) for $|x| < \epsilon$, (X, Y) in a δ-neighborhood of zero, then

$$(5.5)\qquad |\psi(x, X, Y)| \leqslant K|x|.$$

Also, $K(\epsilon, \delta) \longrightarrow 0$ as $(\epsilon, \delta) \longrightarrow (0, 0)$. Thus, in a neighborhood of $(x, y) = 0$, $(X, Y) = 0$, every periodic orbit of (5.2) must have the y coordinate satisfying (5.5). For $x = (x_1, x_2)$, this justifies the transformation of variables

$$(5.6)\qquad x_1 = u \cos \theta, \quad x_2 = -u \sin \theta, \quad y = uv$$

in (5.2). The new equations for (θ, u, v) are

$$\dot{\theta} = 1 - (\widetilde{X}_1 \sin\theta + \widetilde{X}_2 \cos\theta)/u \overset{\text{def}}{=} 1 + \Theta(\theta, u, v, X, Y),$$

$$(5.7) \qquad \dot{u} = \widetilde{X}_1 \cos\theta - \widetilde{X}_2 \sin\theta,$$

$$\dot{v} = Av + \widetilde{Y}/u - (\widetilde{X}_1 \cos\theta - \widetilde{X}_2 \sin\theta)v/u$$

where $\widetilde{X} = (\widetilde{X}_1, \widetilde{X}_2)$, $\widetilde{X} = X_0 + X$, $\widetilde{Y} = Y_0 + Y$ and all functions are evaluated at $(u\cos\theta, -u\sin\theta, uv)$.

Since the function Θ satisfies $\Theta(\theta, 0, v, 0, 0) = 0$, it follows that $\dot{\theta} > \frac{1}{2}$ for v in a fixed compact set and (u, X, Y) in a sufficiently small neighborhood of $(0, 0, 0)$. Thus, we may replace t by θ to obtain

$$(5.8) \qquad du/d\theta = f(\theta, u, v, X, Y), \qquad dv/d\theta = Av + g(\theta, u, v, X, Y).$$

The functions f, g are 2π-periodic in θ and

$$(5.9) \qquad f(\theta, 0, v, X, Y) = 0, \quad \partial f(\theta, 0, v, 0, 0)/\partial u = 0, \quad g(\theta, 0, v, 0, 0) = 0.$$

Any 2π-periodic solution of equation (5.8) corresponds to a periodic orbit of (5.2) through the transformation (5.6) and conversely.

For equation (5.8), there is the standard procedure of alternative problems or Liapunov-Schmidt for determining the 2π-periodic solutions for (u, X, Y) in a neighborhood of zero and v in a compact set (see Cesari [1] for a general discussion of the alternative method as well as references). To describe the method, let $P_{2\pi} = \{w: \mathbf{R} \longrightarrow \mathbf{R}^{n+1}, 2\pi\text{-periodic, continuous}\}$ with the supremum topology and, for $\epsilon > 0$, let $W(\epsilon)$ be the ϵ-neighborhood of $(X, Y) = 0$. For $a \in \mathbf{R}$ fixed, consider the equation, for $(u, v) \in P_{2\pi}$,

$$\frac{1}{2\pi}\int_0^{2\pi} u(s)ds = a,$$

$$(5.10) \qquad \frac{du}{d\theta} = f(\theta, u, v, X, Y) - \frac{1}{2\pi}\int_0^{2\pi} f(s, u(s), v(s), X, Y)\,ds,$$

$$\frac{dv}{d\theta} = Bv + g(\theta, u, v, X, Y).$$

An application of the Implicit Function Theorem shows there are an $\epsilon > 0$ and a unique C^k-function $(u(\cdot, a, X, Y), v(\cdot, a, X, Y)) \in P_{2\pi}$ for $|a| < \epsilon$, $(X, Y) \in W(\epsilon)$, vanishing for $(a, X, Y) = (0, 0, 0)$ and satisfying (5.10). If we define

$$(5.11) \qquad G(a, X, Y) = \frac{1}{2\pi}\int_0^{2\pi} f(s, u(s, a, X, Y), v(s, a, X, Y), X, Y)\,ds$$

then the above 2π-periodic function will satisfy (5.8) if and only if

$$(5.12) \qquad G(a, X, Y) = 0.$$

It is also easy to show that every 2π-periodic solution of (5.8) for (u, X, Y) in a neighborhood of zero and v in a compact set can be obtained through this process.

The function $G(a, X, Y)$ is called the *bifurcation function* and equation (5.12) is called the *bifurcation equation.*

We can now prove the following result. The conclusions on stability are due to deOliveira and Hale [1].

THEOREM 5.2. *There are a neighborhood U of zero in $P_{2\pi}$, an $\epsilon > 0$ and C^k-functions*

$$u: \mathbf{R} \times V(\epsilon) \times W(\epsilon) \longrightarrow \mathbf{R},$$

$$v: \mathbf{R} \times V(\epsilon) \times W(\epsilon) \longrightarrow \mathbf{R}^n,$$

$$G: V(\epsilon) \times W(\epsilon) \longrightarrow \mathbf{R},$$

$V(\epsilon) = (-\epsilon, \epsilon)$, $W(\epsilon)$ *the ϵ-neighborhood of* $(X, Y) = (0, 0)$, $(u(\cdot, a, X, Y), v(\cdot, a, X, Y))$ $\in P_{2\pi}$ *such that equation* (5.8) *has a 2π-periodic solution* $(u(\theta), v(\theta))$ *in U if and only if* $(u(\theta), v(\theta)) = (u(\theta, a, X, Y), v(\theta, a, X, Y))$ *and a satisfies* (5.12). *Furthermore, the stability properties of a 2π-periodic solution $u(\theta, a_0, X, Y)$, $v(\theta, a_0, X, Y)$ coincide with the stability properties of a_0 as a solution of the scalar equation*

$$\dot{a} = G(a, X, Y). \tag{5.13}$$

In particular, $u(\theta, a, X, Y)$, $v(\theta, a, X, Y)$ is stable (asymptotically stable) (unstable) if and only if a_0 is stable (asymptotically stable) (unstable).

PROOF. We only outline the proof of stability. The details can be found in deOliveira and Hale [1] or Hale [2]. The theorem is first proved for the variable v absent; that is, a scalar equation

$$du/d\theta = \tilde{f}(\theta, u, X, Y). \tag{5.14}$$

Let $\tilde{u}(\theta, a, X, Y)$, $\tilde{G}(a, X, Y)$ be the 2π-periodic function and bifurcation function constructed for this equation by the alternative method. Then $\partial\tilde{u}(\theta, a, X, Y)/\partial a = 1$ at $a = 0$, $(X, Y) = (0, 0)$ and one can make the transformation of variables $u \mapsto b$, $\tilde{u} = u(\theta, b, X, Y)$ to obtain

$$\dot{b} = \left[\frac{\partial\tilde{u}(\theta, b, X, Y)}{\partial b}\right]^{-1} \tilde{G}(b, X, Y).$$

Since $\partial\tilde{u}/\partial b > 0$ if b, X are small, one obtains the result for v absent. When v is present, there is a center manifold with the flow on the center manifold given by an equation of the form (5.14). One now has two bifurcation functions $\tilde{G}(a, X, Y)$ and $G(a, X, Y)$ corresponding respectively to (5.14) and (5.8). These functions must have the same zeros and one can use an argument similar to the one in the proof of Theorem 4.2.

Theorem 5.2 has obvious interpretations about the existence and stability of periodic orbits of (5.2). Using the same type of argument that was used in the proof of Theorem 4.2, one can also show that the ω-limit set of an orbit in a neighborhood of zero consists of either zero or a single periodic orbit. These results are summarized in

THEOREM 5.3. *Let $u(\theta, a, X, Y)$, $v(\theta, a, X, Y)$ be the 2π-periodic functions satisfying* (5.10) *and let $G(a, X, Y)$ be defined in* (5.11). *Then there are a neighborhood V of $(x, y) = (0, 0)$ and a neighborhood W of $(X, Y) = (0, 0)$ such that equation* (5.2) *for (X, Y) in W has a periodic solution $(x(t), y(t))$ in V if and only if*

$$x(t) = (u(\theta(t), a_0, X, Y)\cos\theta(t), -u(\theta(t), a_0, X, Y)\sin\theta(t)),$$

$$y(t) = u(\theta(t), a_0, X, Y)v(\theta(t), a_0, X, Y)$$

where $G(a_0, X, Y) = 0$, $\dot\theta(t) = 1 + \Theta(\theta, a_0, X, Y), v(\theta, a_0, X, Y))$ and $\dot\theta(0)$, a_0 are uniquely determined by $x(0)$, $y(0)$. The stability properties of the periodic orbit coincide with the stability properties of a_0 as a solution of (5.13). *Finally, the ω-limit set of an orbit of* (5.2) *in V is a single periodic orbit.*

One can show (see, for example, Chafee [1]) that

$$G(a, X, Y) = -G(-a, X, Y). \tag{5.15}$$

Thus, if a is a solution of $G(a, X, Y) = 0$, then so is $-a$. One shows that these two solutions will correspond to the same periodic orbit of (5.2). Thus, we only need to be concerned with positive roots of $G(a, X, Y) = 0$.

Using this fact, Theorem 5.2 and the ideas of the proof of Corollary 4.4, one obtains the following interesting consequence for the stability of the zero solution of the unperturbed equation (5.1).

COROLLARY 5.4. *The zero solution of* (5.1) *is asymptotically stable (unstable) if and only if there is an $\epsilon > 0$ such that $aG(a, 0, 0) < 0$ (> 0) for $0 < |a| < \epsilon$. There is an $\epsilon > 0$ such that $G(a, 0, 0) = 0$ for $|a| < \epsilon$ if and only if the zero solution is stable and there is a first integral in a neighborhood of zero.*

This corollary was first proved by Liapunov [1] (see also Bibikov [1]) for the case when X_0, Y_0 are analytic. In particular, Corollary 5.4 implies that, if

$$G(a, 0, 0) = \beta_0 a^{2q+1} + o(|a|^{2q+1}) \quad \text{as } |a| \longrightarrow 0,\ \beta_0 \neq 0, \tag{5.16}$$

then the zero solution of (5.1) is asymptotically stable if $\beta_0 < 0$ and unstable if $\beta_0 > 0$. By using the theory of normal forms for equation (5.1) (see, for example, Bibikov [1] or Takens [1]) one can relate the sign of β_0 to the stability of equation (5.1) under certain types of high order perturbation (for details, see Negrini and Salvadori [1], Bernfeld and Salvadori [1]).

For $k \geqslant 2q + 1$, we now discuss equation (5.2) with $G(a, 0, 0)$ satisfying (5.16). There are functions α_j: $W \longrightarrow \mathbf{R}$, $\alpha_j(0, 0) = 0$, $0 \leqslant j \leqslant q - 1$, $\alpha_q(0, 0) = \beta_0$ such that

$$G(a, X, Y) = \sum_{j=0}^{q} \alpha_j(X, Y)a^{2j+1} + o(|a|^{2q+1}) \tag{5.17}$$

as $|a| \longrightarrow 0$. Let

$$\text{(5.18)}\qquad \begin{aligned} &\Gamma^0 = \{(X, Y): \alpha_0(X, Y) = 0\}, \quad \Gamma^+ = \{(X, Y): \alpha_0(X, Y) > 0\}, \\ &\Gamma^- = \{(X, Y): \alpha_0(X, Y) < 0\}. \end{aligned}$$

All of the coefficients $\alpha_j(X, Y)$ can be computed in terms of the derivatives of X, Y of order $\leqslant 2j + 1$. For the matrix of the linear approximation of (5.2) at zero, the function $\alpha_0(X, Y)$ represents the real parts of the eigenvalues which are purely imaginary for $(\widetilde{X}, \widetilde{Y}) = (0, 0)$. Thus, in order to have a periodic orbit bifurcate from zero by a variation in $(\widetilde{X}, \widetilde{Y})$ near some point $(\widetilde{X}, \widetilde{Y})$, the point $(\widetilde{X}, \widetilde{Y})$ must have $\alpha_0(\widetilde{X}, \widetilde{Y}) = 0$; that is $(\widetilde{X}, \widetilde{Y})$ belongs to Γ^0.

For some special types of one parameter families of vector fields, the bifurcation at (X_0, Y_0) is very simple as shown in the following result, essentially due to Negrini and Salvadore [1].

THEOREM 5.5. *Suppose $G(a, 0, 0)$ satisfies (5.16), $(X(\mu), Y(\mu))$, $\mu \in \mathbf{R}$, is a C^1-family of vector fields, vanishing at $\mu = 0$, and suppose the functions $\alpha_j(X(\mu), Y(\mu)) \overset{\text{def}}{=} \alpha_j(\mu)$ are defined in (5.17). If $\alpha_0'(0) \neq 0$, then there are a neighborhood V of $(x, y) = (0, 0)$ and a $\mu_0 > 0$ such that, if $0 < |\mu| < \mu_0$, then equation (4.2) has a periodic orbit in V if and only if $\alpha_0'(0)\beta_0 < 0$. When this condition is satisfied the orbit is unique. It is asymptotically stable if $\beta_0 < 0$ and unstable if $\beta_0 > 0$.*

PROOF. The bifurcation function $G(a, X(\mu), Y(\mu)) = G(a, \mu)$ satisfies

$$G(a, \mu) = \sum_{j=0}^{q} \alpha_j(\mu) a^{2j+1} + o(|a|^{2q+1}) \quad \text{as } |a| \longrightarrow 0$$

where $\alpha_j(0) = 0$, $0 \leqslant j < q$, $\alpha_q(0) = \beta_0 \neq 0$. It is not difficult to show that there is a neighborhood U of $(a, \mu) = (0, 0)$ such that any solution of $G(a, \mu) = 0$ in U has the a priori bound $|a| \leqslant k|\mu|^{1/2q}$. If we let $a = |\mu|^{1/2q}b$, then

$$\frac{1}{|\mu|}(G(|\mu|^{1/2q}b, \mu) = F(b, \mu) \overset{\text{def}}{=} \sum_{j=0}^{q} \gamma_j(\mu) b^{2j+1} + o(|\mu|)$$

as $|\mu| \longrightarrow 0$, where $\gamma_0(0) = \alpha_0'(0)$, $\gamma_j(0) = 0$, $0 < j < q$, $\gamma_q(0) = \beta_0 \neq 0$. The Implicit Function Theorem implies the result stated in the theorem on existence. The stability follows from Theorem 5.2.

Theorem 5.5 says that nothing very complicated can occur from variations in a one parameter family of vector fields regardless of the degeneracy in $G(a, 0, 0)$ provided that the eigenvalues cross the imaginary axis with a definite speed; that is, $\alpha_0(\mu) = \alpha_0'(0)\mu + o(|\mu|)$ as $\mu \longrightarrow 0$, $\alpha_0'(0) \neq 0$. It is only possible to obtain one periodic orbit and the amplitude of the orbit (from the proof) is of order $|\mu|^{1/2q}$ as $|\mu| \longrightarrow 0$.

Theorem 5.5 does not imply that there is at most one periodic orbit of (5.2) for every (X, Y) in a neighborhood of $(0, 0)$. In fact, if $q > 1$, we will see later that there will always be some (X, Y) in any neighborhood of zero such that there are q periodic orbits. Assuming one parameter families of perturbations in (5.2), Flockerzi [1] has used Newton's polygon to determine the number of solutions of $G(a, \mu) = 0$ and their stability when $\alpha_0(\mu) = \alpha_0\mu^k + o(|\mu|^k)$ as $|\mu| \longrightarrow 0$. We obtain results for arbitrary perturbations (X, Y) which are more in the spirit of Chafee [1].

The first result for $q = 1$ is referred to as the *generic Hopf Bifurcation Theorem.*

THEOREM 5.6. *Suppose*

$$G(a, 0, 0) = \beta_0 a^3 + o(|a|^3) \quad \textit{as } |a| \longrightarrow 0, \ \beta_0 \neq 0,$$

and $\alpha_0(X, Y)$ *is defined in* (5.17). *Then there are neighborhoods* W *of* $(X, Y) = (0, 0)$, V *of* $(x, y) = (0, 0)$ *such that for* (X, Y) *in* W, *equation* (5.2) *has a periodic orbit in* V *if and only if* $\alpha_0(X, Y)\beta_0 < 0$. *When this condition is satisfied, the orbit is unique and is asymptotically stable (unstable) if and only if* $\beta_0 < 0$ (> 0).

PROOF. In a neighborhood of zero, $G(x, X, Y) = 0$ either has one positive root or no positive root. The condition for the existence of one is $\alpha_0(X, Y)\beta_0 < 0$. The stability of the orbit follows from Theorem 5.2.

With the notation as in (5.18), if $(X, Y) \in \Gamma^+(X, Y)$, then a periodic orbit can exist for (X, Y) in $\Gamma^+(X, Y)$ if and only if $\beta_0 < 0$ and then it is asymptotically stable (the *supercritical case*) with the origin being unstable. It can exist for $(X, Y) \in \Gamma^-$ if and only if $\beta_0 > 0$ and then it is unstable (the *subcritical case*) with the origin being stable.

Suppose there is a one parameter family of perturbations $X(\lambda)$, $Y(\lambda)$, $\lambda \in \mathbf{R}$, which vanish for $\lambda = 0$ and this curve of perturbations has $(X(\lambda), Y(\lambda))$ in Γ^- for $-\epsilon < \lambda < 0$ and in Γ^+ for $0 < \lambda < \epsilon$. Then the origin is asymptotically stable for $\lambda < 0$ and unstable for $\lambda > 0$. There are an unstable periodic orbit and subcritical bifurcation at $\lambda = 0$ if $\beta_0 > 0$ and an asymptotically stable periodic orbit and supercritical bifurcation at $\lambda = 0$ if $\beta_0 < 0$. Notice that it is not required that the curve cross Γ^0 transversally; that is, the eigenvalues of the linear part of (5.2) cross the imaginary axis with a definite speed. Contrast this with Theorem 5.5.

If the first, second and third derivatives of $G(a, 0, 0)$ vanish at $a = 0$, Theorem 5.6 cannot be used. The discussion of the case of higher order degeneracy of $G(a, 0, 0)$ is much more complicated and one is forced to use the Malgrange Preparation Theorem on $G(a, X, Y)$ for the general case as we did in §4. There are some more specific results presented below that can be obtained without this theorem; namely the case $q = 2$ in (5.16).

Suppose $q = 2$ in (5.16) and define $\Gamma^0, \Gamma^+, \Gamma^-$ by (5.18). If $\alpha_0(X, Y)\beta_0 < 0$, then $\alpha_0(X, Y)\alpha_3(X, Y) < 0$ for (X, Y) in a neighborhood of zero and $G(a, X, Y)$ has exactly one positive zero. Thus, equation (5.2) has exactly one periodic orbit. If $\alpha_0(X, Y)\beta_0 \geq 0$, there may be more than one positive zero of $G(a, X, Y)$. To determine conditions for when there are more than one, define $P(r, X, Y)$, $r \geq 0$, by the relation

$$G(a, X, Y) = aP(a^2, X, Y). \tag{5.19}$$

Extend $P(r, X, Y)$ as an even function for $r \in \mathbf{R}$. Then the positive zeros of G are determined from the positive zeros of $P(r, X, Y)$. Since $\beta_0 \neq 0$, the Implicit Function Theorem implies there are neighborhoods V of $r = 0$, W of $(X, Y) = 0$ and a unique function δ: $W \longrightarrow \mathbf{R}^+$ such that $\delta(0, 0) = 0$ and $\partial P(\delta(X, Y), X, Y)/\partial r = 0$. Let

$$\gamma(X, Y) = P(\delta(X, Y), X, Y). \tag{5.20}$$

If $\gamma(X, Y)\beta_0 < 0$, there are two positive simple zeros of $P(r, X, Y)$ and thus two periodic orbits of equation (5.2). The orbit corresponding to the smaller zero is unstable (asymptotically stable) if $\beta_0 < 0$ (> 0) and the other is asymptotically stable (unstable). If $\gamma(X, Y)\beta_0 = 0$, $\delta(X, Y) > 0$, then there is a unique periodic orbit which on a center manifold is stable on one side and unstable on the other. If $\gamma(X, Y)\beta_0 > 0$, there is no positive zero of $P(r, X, Y)$ and, thus, no periodic orbit of (5.2). These results are summarized in

THEOREM 5.7. *Suppose*

$$G(a, 0, 0) = \beta_0 a^5 + o(|a|^5) \quad \textit{as } |a| \longrightarrow 0,\ \beta_0 \neq 0,$$

$\alpha_0(X, Y)$ *is defined in* (5.17) *and* $\gamma(X, Y)$ *is defined by* (5.20) *when* $\alpha_0(X, Y)\beta_0 \geq 0$. *Then there are neighborhoods* W *of* $(X, Y) = (0, 0)$, V *of* $(x, y) = (0, 0)$ *such that, for* $(X, Y) \in W$, *equation* (5.2) *satisfies the following properties in* V:

(i) $\alpha_0(X, Y)\beta_0 < 0$ *implies a unique periodic orbit which is asymptotically stable (unstable) if* $\beta_0 < 0$ (> 0).

(ii) $\alpha_0(X, Y)\beta_0 \geq 0$, $\gamma(X, Y)\beta_0 < 0$ *implies there are two periodic orbits, the one corresponding to the smaller value of a is unstable (asymptotically stable) with the other one being asymptotically stable (unstable) if* $\beta_0 < 0$ (> 0).

(iii) $\alpha_0(X, Y)\beta_0 \leq 0$, $\gamma(X, Y) = 0$, $\delta(X, Y) > 0$ *implies there is a unique periodic orbit which on a center manifold is stable on one side and unstable on the other.*

(iv) $\alpha_0(X, Y)\beta_0 \geq 0$, $\gamma(X, Y)\beta_0 > 0$ *implies no periodic orbit.*

If we let $\Sigma = \{(X, Y): \delta(X, Y) = 0\}$, Σ has codimension 1 and is tangent to Γ^0 at $(0, 0)$. Define Γ^0 as in (5.18). Then Theorem 5.7 states that the bifurcation set for equation (5.2) for $q = 2$ in (5.16) is given by $\Gamma^0 \cup \Sigma$. The neighborhood W is divided into the components shown in Figure 4 with the number of periodic orbits indicated. The set Γ^+ is drawn to the right of Γ^0 and the intersection of Γ^0, Σ contains the point $(X, Y) = (0, 0)$.

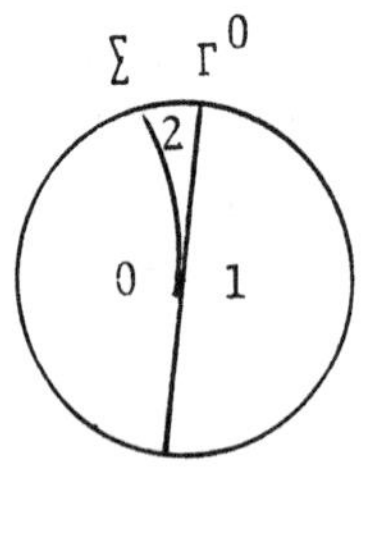

FIGURE 4a. $\beta_0 < 0$.

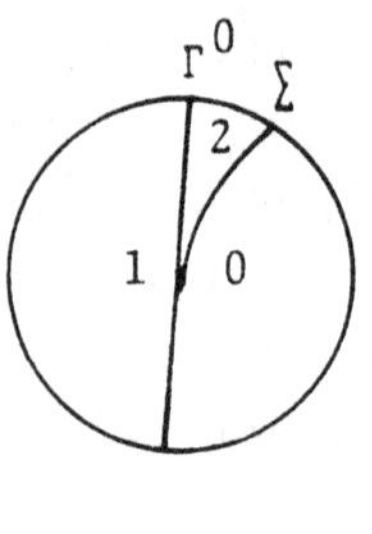

FIGURE 4b. $\beta_0 > 0$.

From these pictures, one can see clearly the meaning of Theorem 5.7 for $q = 2$. Any curve of perturbations $(X(\mu), Y(\mu))$ crossing Γ^0 transversally, that is, $\Gamma'(0) \neq 0$, can never enter the region containing two orbits. If the curve is not transversal to Γ^0, it is possible for it to enter this region. As remarked earlier, this type of problem with one parameter problems was discussed by Flockerzi [1].

There remains the problem of computing the number β_0 in (5.16) and the functions $\alpha_j(X, Y)$ in (5.17). One method is to approximate the Taylor series in a for the functions $u(a, X, Y)$, $v(a, X, Y)$, $G(a, X, Y)$ directly from the defining equations (5.10), (5.11). An alternative procedure is to transform the vector field (5.2) by a change of coordinates to a convenient normal form. The method of Lie transforms developed originally for applications to celestial mechanics is very convenient for obtaining this normal form. The best reference is Henrard [1]. This method also is presented in the forthcoming book of Chow and Hale [1]. It leads naturally to implementation on the computer (see Meyer and Schmidt [1]) and could lead to considerable improvement over the present methods used for the computation of periodic orbits.

Using the same ideas, the appropriate extensions of the results to the case where A has eigenvalues with nonzero real parts are only a technicality.

In the applications, it often happens that the perturbed vector field depends in a special way on a finite number of parameters varying in a neighborhood of zero. The bifurcation diagrams in the parameter space are then obtained by studying the way in which the family of vector fields crosses the general bifurcation curves given above. For example, if there is a single parameter μ and the lowest order terms in the bifurcation function are $\beta a^3 + \gamma\lambda$ with $\beta \neq 0$, $\gamma \neq 0$, then this is the case of generic Hopf bifurcation, and the dependence on λ through the linear term $\gamma\lambda$ implies there is only one crossing of the bifurcation curve as we noted before. When there are several parameters in the generic Hopf bifurcation problem, then the family of vector fields can cross several times and in a very complicated manner. The case where the degeneracy in a is of order 5 ($q = 2$ in (5.16)) is even more difficult. Also, it is often the case that one parameter, say μ, plays a distinguished role (sometimes called a modal parameter) and one is interested in studying the bifurcations when this parameter is fixed and the other parameters are varied. This question has been discussed in detail for third order and fifth order degeneracy by Golubitsky and Langford [1].

6. First order bifurcation in the plane

Suppose $\Omega \subset \mathbf{R}^2$ is open and let $\mathcal{X}_2^r = \mathcal{X}_2^r(\overline{\Omega})$ be the set of $C^r(\overline{\Omega}, \mathbf{R}^2)$ vector fields, $r \geqslant 1$, which are transverse to the boundary of Ω. The set Σ_2^r of structurally stable vector fields in $\mathcal{X}_2^r$ is characterized in Theorem 3.2 and Σ_2^r is open and dense in $\mathcal{X}_2^r$. Any $f \in \mathcal{X}_2^r$ which is not structurally stable is a *bifurcation point.* Our objective in this section is to discuss some of the most elementary bifurcation points.

Following Andronov and Leontovich [1] (see also Andronov et al. [1]) except for terminology, we say $f \in \mathcal{X}_2^r$ is a *bifurcation point of degree* 0 if it is structurally stable. It is a *bifurcation point of degree* 1 if it is not of degree zero and every g in a neighborhood of f is either of degree zero or equivalent to f. It is a *bifurcation point of degree* 2 if it is not of degree 0 or 1 and every g in a neighborhood of f is either of degree 0 or 1 or equivalent to f. Similarly, one defines a *bifurcation point of degree k.*

The bifurcation points of degree one are the simplest types that can occur. They correspond to those vector fields which satisfy all of the conditions for structural stability except one and the condition that is violated must be done so in the simplest way. To be more specific, suppose an equilibrium point $x_0 = 0$ of f is not hyperbolic. This can occur only when $A = \partial f(0)/\partial x$ has an eigenvalue on the imaginary axis. Now consider all g near f which have an equilibrium point near zero which is not hyperbolic. In order for all such g to be equivalent to f, it is necessary that A has either a simple zero or a pair of simple purely imaginary eigenvalues. If A has a simple zero and f has degree 1 in $\mathcal{X}_2^r$, $r \geqslant 2$, then the bifurcation function $G(a, f)$ in §4 for equilibrium points near $x_0 = 0$ must have the form $G(a, f) = \beta_0 a^2 + o(|a|)$ as $|a| \to 0$, $\beta_0 \neq 0$, and Theorem 4.8 is applicable. There are a neighborhood W of f and a submanifold Γ of codimension one such that $W \backslash \Gamma = U_1 \cup U_2$ where $g \in U_1$ implies no equilibrium point near x_0, $g \in U_2$ implies a saddle and hyperbolic node near x_0 and $g \in \Gamma$ implies a saddle-node near x_0, which we call an *elementary saddle-node.*

If A has a pair of purely imaginary roots and f has degree 1 in $\mathcal{X}_2^r$, $r \geqslant 3$, then the bifurcation function $G(a, f)$ in §5 for the existence of periodic orbits near x_0 must have the form

$$G(a, f) = \beta_0 a^3 + o(|a|^3) \quad \text{as } |a| \to 0,\ \beta_0 \neq 0.$$

This corresponds to the generic Hopf bifurcation (Theorem 5.6). There are a neighborhood W of f and a submanifold Γ of codimension one such that $W \backslash \Gamma = U_1 \cup U_2$ and $g \in U_1$

implies no periodic orbit near x_0 and $g \in U_2$ implies a unique periodic orbit near x_0. We refer to the equilibrium point of g near x_0 as an *elementary focus* if $g \in \Gamma$.

The condition for f to be a bifurcation point also can occur when a periodic orbit γ becomes not hyperbolic. Let $\gamma = \{\phi(t, p), t \in \mathbf{R}\}$ where $\phi(t, p)$, $\phi(0, p) = p$, is a periodic solution of $\dot{x} = f(x)$. Let L_0 be a transversal to γ at p and π: $L_0 \longrightarrow L_0$ be the Poincaré map. Then $\pi(p) = p$ and not hyperbolic implies $\pi'(p) = 1$. If f has degree one in $\mathcal{X}_2^r$, $r \geqslant 2$, then it must necessarily be true that $\pi''(p) \neq 0$. This means that $\pi(p + u) - (p + u)$ in a neighborhood of $u = 0$ behaves like a quadratic function in u. Then there will be a neighborhood W of f and a submanifold Γ of codimension one such that $W \backslash \Gamma = U_1 \cup U_2$ and $g \in U_1$ implies no periodic orbits near γ (no fixed points of π near p) and $g \in U_2$ implies two hyperbolic periodic orbits near γ (two fixed points of π near p), one unstable and the other asymptotically stable. If $g \in \Gamma$, then there is a unique periodic orbit near γ, stable on one side and unstable on the other. Having $g \in \Gamma$ is equivalent to having a double zero of $\pi(p + u) - (p + u)$ near $u = 0$. This result on periodic orbits can also be obtained from the method in §4 by introducing a local coordinate system $x \mapsto (\theta, \rho)$, $x = \phi(\theta, p) + \rho v(\theta)$, where $v(\theta)$ is a unit vector orthogonal to $\partial\phi(\theta, p)/\partial\theta$. If t is replaced by θ in the new equations, one obtains a scalar equation for ρ as a periodic function of θ, a special case of equation (5.8). Theorem 5.2 is then applicable.

If the equilibrium points and periodic orbits are hyperbolic and f is a bifurcation point, then there must be a trajectory connecting saddle points. However, if f has degree one, then one can show that there must be an orbit whose α- and ω-limit sets are the same point—a *homoclinic orbit.* This situation is more complicated to understand than the previous ones because it is a global problem for which the discussion cannot be restricted to the consideration of only the fixed points of a map.

To describe the behavior near a homoclinic orbit, suppose zero is a saddle point of f, $f(0) = 0$, $\partial f(0)/\partial x$ has eigenvalues $\lambda < 0 < \mu$. If W_f^s, W_f^u are the stable and unstable manifolds of 0, then there is a homoclinic orbit through zero if and only if $(W_f^s \cap W_f^u) \backslash \{0\} \neq \emptyset$. If $p_0 \in W_f^s \cap W_f^u$, $p_0 \neq 0$, then the orbit $O(p_0)$ through p_0 approaches zero exponentially as $t \longrightarrow \pm\infty$. If $\gamma = O(p_0) \cup \{0\}$, then the invariant set γ can have either of the configurations shown in Figure 5 with respect to the position of the stable and unstable manifolds of zero. In order to be specific, we suppose the situation in Figure 5a occurs. The other case is discussed in a similar way.

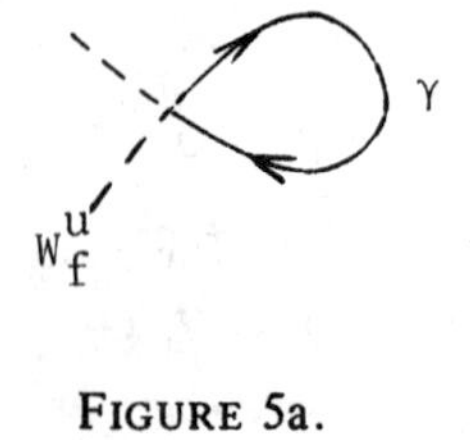

FIGURE 5a.

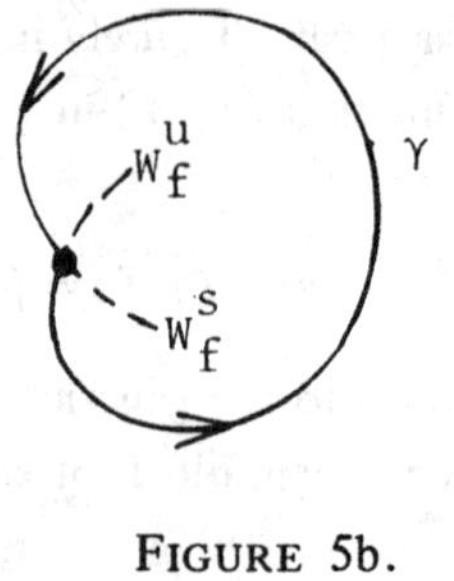

FIGURE 5b.

Let L_0 be a transversal to γ at p_0. Let L_0^+ (L_0^-) be that part of L_0 interior (exterior) to γ in Figure 6. There is a neighborhood U of p_0 such that, for any $p \in U \cap L_0^+$, there is a $\tau = \tau(p) > 0$ with $x(\tau(p), p) \in L_0^+$, $x(t, p) \notin L_0^+$, $0 < t < \tau(p)$, where $x(t, p)$ is the solution of $\dot{x} = f(x)$ through p. If $\pi_0(p) = x(\tau(p), p)$, then, for any integer k, positive or negative, one can define $\pi_0^k(p)$ on some subset of L_0^+ depending on k. The map π_0 is like a Poincaré map relative to the interior of γ. It is not defined at p_0 and $\tau(p) \to \infty$ as $p \to p_0$.

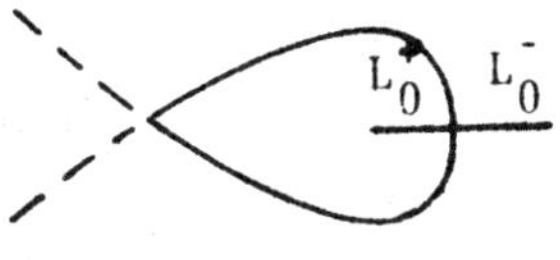

FIGURE 6.

DEFINITION 6.1. *The homoclinic orbit is asymptotically stable* (*unstable*) *if* $\pi_0^k(p) \to p_0$ *as* $k \to \infty$ ($k \to -\infty$). *It is exponentially asymptotically stable* (*exponentially unstable*) *if the approach to* p_0 *is exponential.*

One can show that the stability (or instability) of γ is exponential if and only if $\sigma_0 = \operatorname{tr} \partial f(0)/\partial x = \lambda + \mu < 0$ (or > 0) (see Andronov et al. [1], Chow and Hale [1]).

There is a neighborhood W of f such that each $g \in W$ has a unique zero near $x = 0$. By translation of variables, we can assume $g(0) = 0$. Let W_g^s, W_g^u be the stable and unstable manifolds of zero. Since these manifolds are smooth in g, there are points $p_{u,g}$ of first contact with L_0 ($p_{s,g}$ of last contact with L_0) with respect to increasing t so that one of the pictures in Figure 7 prevails.

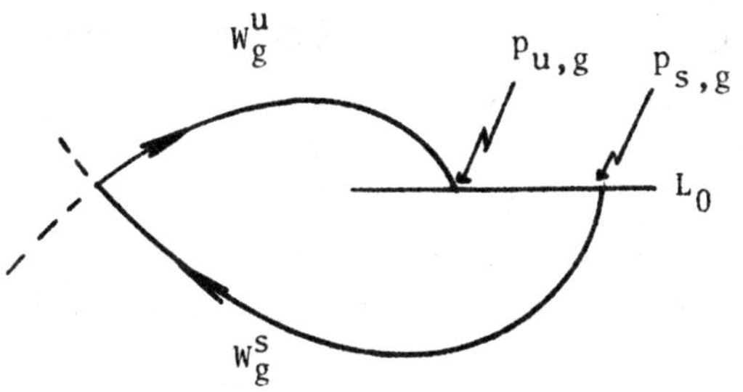

FIGURE 7a.

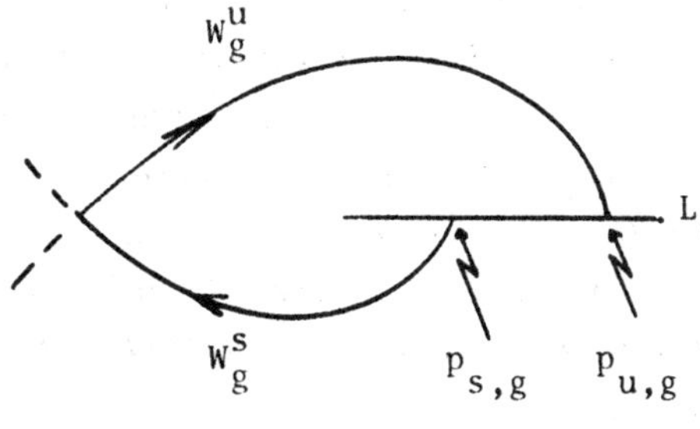

FIGURE 7c

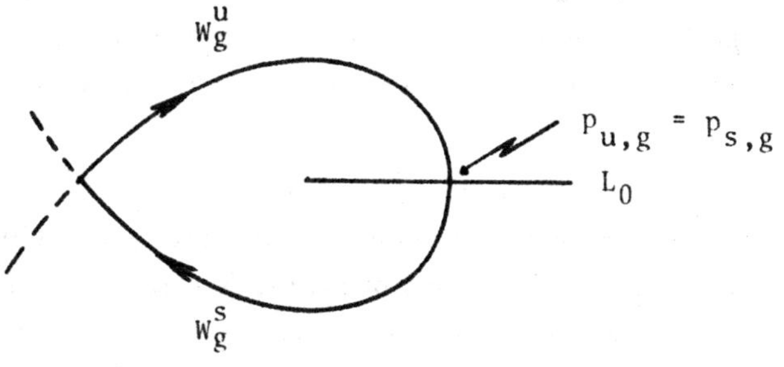

FIGURE 7b.

If case (b) prevails, then g has a homoclinic orbit through 0. If $\Gamma = \{g \in W: p_{u,g} = p_{s,g}\}$, then one can show that Γ has codimension one for a sufficiently small neighborhood W of f. If $W \backslash \Gamma = U_1 \cup U_2$, suppose $g \in U_1$ coincides with Figure 7a and $g \in U_2$ coincides with Figure 7c. If $g \in U_1$, then there is a neighborhood V of γ containing only the critical point zero, such that every solution with initial value on I_g, the segment between $p_{u,g}$ and $p_{s,g}$, leaves V in negative time. Thus, the behavior of solutions near γ is determined by what happens to the positive orbit $O^+(p_{u,g})$ of $p_{u,g}$. If γ is asymptotically stable, then $O^+(p_{u,g})$ remains in V and the Poincaré-Bendixson Theorem implies there is a periodic orbit in V. If it is further assumed that γ is exponentially asymptotically stable, then there is a unique periodic orbit in V (see Andronov et al. [1] or Chow and Hale [1]). If γ is exponentially unstable, then $O^+(p_{u,g})$ leaves V in finite time.

If $g \in U_2$, that is, Figure 7c prevails, then the behavior of solutions near γ is determined by the negative orbit $O^-(p_{s,g})$ of $p_{s,g}$. If γ is unstable, then there is a periodic orbit in V and it is unique if γ is exponentially unstable. If γ is exponentially asymptotically stable, then $O^-(p_{u,g})$ leaves V in finite time.

Finally, one can show that f having hyperbolic equilibrium points and periodic orbits is a bifurcation point of degree 1 if and only if $\sigma_0 = \operatorname{tr} \partial f(0)/\partial x \neq 0$; that is, the stability or instability of γ is exponential (see Andronov et al. [1]). These results are summarized in

THEOREM 6.2. *A vector field $f \in \mathcal{X}_2^r$, $r \geq 3$, is a bifurcation point of degree 1 if and only if there are a neighborhood W of f and a submanifold Γ of codimension one in W such that $W \backslash \Gamma = U_1 \cup U_2$ where each $g \in U_j$ is structurally stable but $g \not\sim h$ if $g \in U_1$, $h \in U_2$. Then only one of the following situations prevails*:

(i) *$g \in \Gamma$ has an elementary saddle-node at x_0, there are no equilibrium points of g near x_0 if $g \in U_1$ and a saddle and node near x_0 if $g \in U_2$.*

(ii) *$g \in \Gamma$ has an elementary focus at x_0, there is no periodic orbit of g near x_0 if $g \in U_1$ and a periodic orbit near x_0 if $g \in U_2$—the generic Hopf bifurcation.*

(iii) *$g \in \Gamma$ has a periodic orbit γ which is stable from one side, unstable from the other, $g \in U_1$ has no periodic orbit near γ and $g \in U_2$ has two hyperbolic periodic orbits near γ.*

(iv) $\sigma_0 = \operatorname{tr} \partial f(0)/\partial x \neq 0$, *$g \in \Gamma$ has a homoclinic orbit containing a saddle point x_0, $g \in U_1$ has a saddle near x_0 and no periodic orbit near γ, $g \in U_2$ has a saddle point and a unique hyperbolic periodic orbit near γ which coalesce as $g \to \Gamma$.*

(v) *There is a connection between two distinct saddle points.*

Each of the cases (i)–(v) is shown in Figure 8.

There is another interesting phenomenon that can occur with a bifurcation point of degree 1. In case (i), the unstable manifold of the saddle could be connected to the stable manifold of the node to form a closed curve as in Figure 9a. They may then coalesce and disappear to form a hyperbolic periodic orbit.

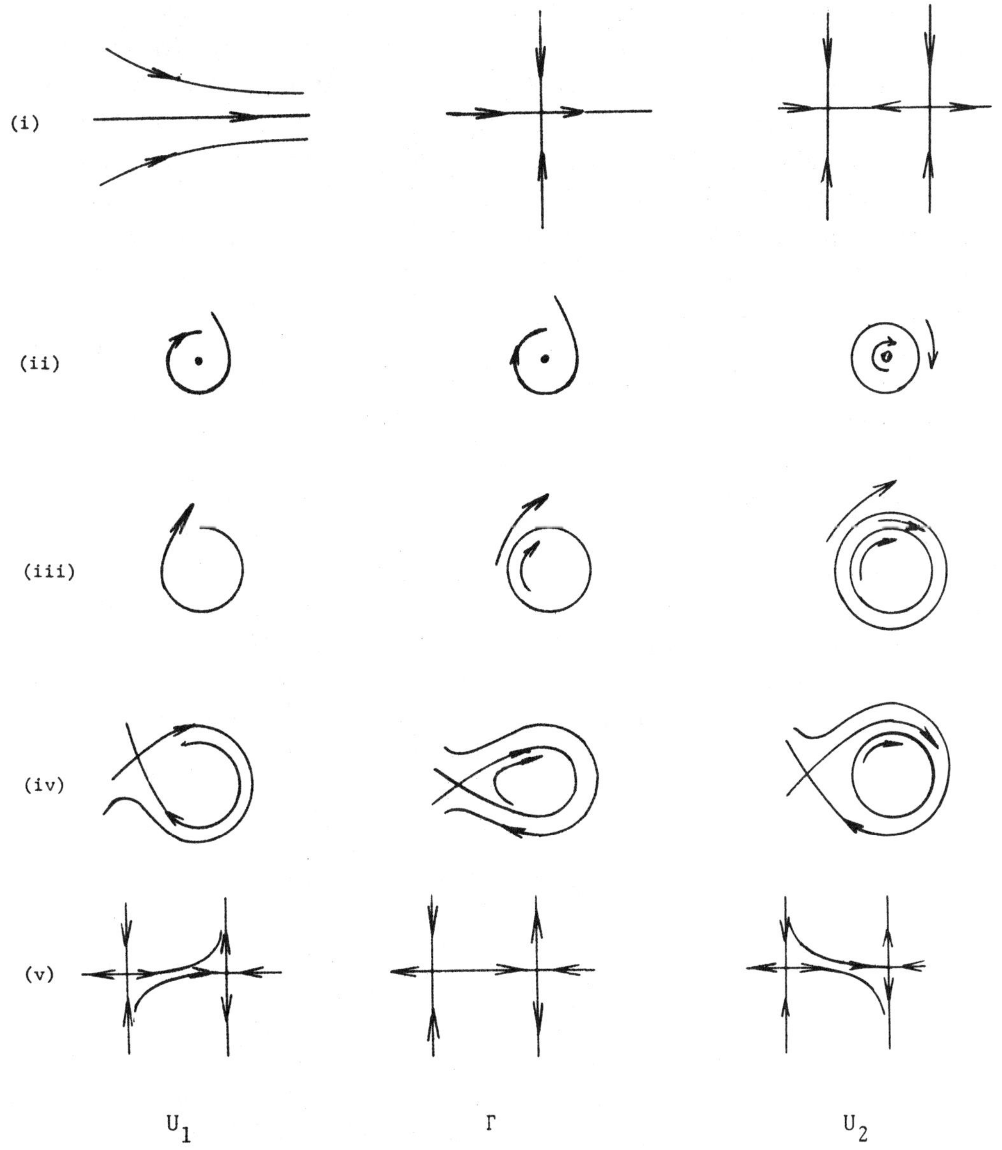

FIGURE 8.

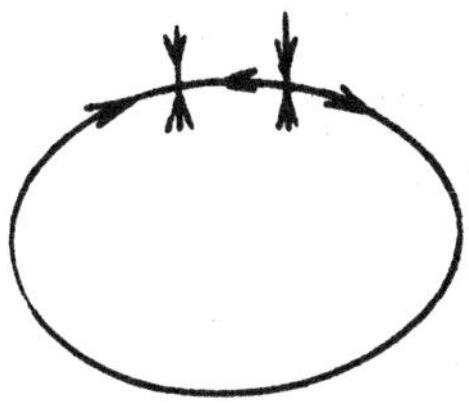

FIGURE 9a.

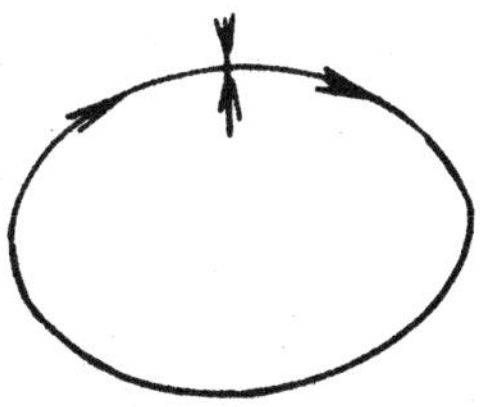

FIGURE 9b.

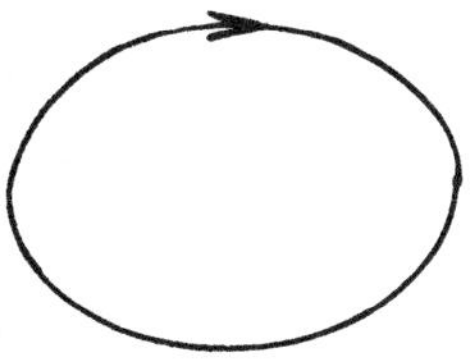

FIGURE 9c.

Using Theorem 6.2, Sotomayor [1] has proved the following interesting result. Let $H = \{g: [0, 1] \longrightarrow X_2^r, g \text{ a } C^r\text{-function}, r \geq 5\}$ with the topology of uniform convergence of the function and its derivatives. Let $H_1 \subset H$ be the subset of functions which have the property that, for any $g \in H_1$, there is a finite set of points $\mu_1, \ldots, \mu_m$ in $[0, 1]$ such that $g(\mu)$ is structurally stable for $\mu \neq \mu_j$, $g(\mu_j)$ is a bifurcation point of degree 1 in X_2^r and the curve defined by g is transversal to the set Γ_j of Theorem 6.2 corresponding to the bifurcation point $g(\mu_j)$. Sotomayor [1] has proved that H_1 is residual in H; that is, one can assume that the bifurcation points on a given curve of vector fields generically will be of the type stated in Theorem 6.2.

Teixeira [1] has generalized the theory of first order bifurcations in the plane to the case where the vector field is required only to be transversal to a disk except at a finite number of points. Newhouse and Peixoto [1] have shown the interesting fact that one can always pass from one Morse-Smale system to another by a one parameter family of vector fields for which every bifurcation point is a saddle-node. This result is valid in dimension n. Generalizations are in Newhouse [2].

The higher order bifurcation points are difficult to classify and only the case of degree 2 has been completely resolved. The local problem near a generalized saddle-node has been discussed in §4 and near a generalized focus in §5. Leontovich [1] has discussed the general case near a homoclinic orbit. For further results, see Andronov et al. [1], Takens [2]. An example will be given later of a higher order bifurcation near an equilibrium point.

7. Two dimensional periodic systems

In this section, our objective is to discuss subharmonic bifurcation, bifurcation of homoclinic orbits and the bifurcation of tori for periodic two dimensional systems. The equation is chosen to have a simple form in order to minimize the notation. Using the same methods, generalizations are possible (see, for example, Chow and Hale [1]).

Suppose $f\colon \mathbf{R} \longrightarrow \mathbf{R}$ is a C^r-function, $r \geqslant 2$, $F\colon \mathbf{R} \longrightarrow \mathbf{R}$ is continuous and periodic of period one. The equation to be considered is

$$\dot{x} = y, \qquad \dot{y} = -f(x) - \mu y + \lambda F(t) \tag{7.1}$$

where $(\lambda, \mu) \in \mathbf{R}^2$ varies in a neighborhood of $(0, 0)$. The unperturbed equation is

$$\dot{x} = y, \qquad \dot{y} = -f(x). \tag{7.2}$$

Suppose equation (7.2) has a periodic orbit Γ_k of least period k, where k is a positive integer.

Problem SB. *Find a neighborhood U of Γ_k and a neighborhood V of* $(\lambda, \mu) = 0$ *such that, for each* $(\lambda, \mu) \in V$, *one knows exactly the number of k-periodic solutions of* (7.1) *which belong to U as well as their stability properties.*

A periodic solution of (7.1) which is k times the period of the forcing function $F(t)$ is called a *subharmonic solution of order k.*

A solution of Problem SB requires some hypotheses. If $\Gamma_k = \{(p_k(t), \dot{p}_k(t)), t \in \mathbf{R}\}$ where $(p_k(t), \dot{p}_k(t))$ is a k-periodic solution of (7.2), then the derivative of this function is a k-periodic solution of the linear variational equation

$$\dot{x} = y, \qquad \dot{y} = -[\partial f(p_k(t))/\partial x]\, y. \tag{7.3}$$

We suppose the set of k-periodic solutions of (7.3) is one dimensional. The periodic orbit Γ_k of the Hamiltonian system (7.2) must belong to a smooth family of periodic orbits $\Gamma_{\omega(a)}$ of period $\omega(a)$ where a is a real parameter in $(-\epsilon, \epsilon)$ with $\omega(0) = k$. One can show the hypothesis on equation (7.3) is equivalent to $\omega'(0) \neq 0$. This hypothesis is generic with respect to the class of vector fields $f \in C^2(\mathbf{R}, \mathbf{R})$, $n \geqslant 2$, with the Whitney topology. Also, if Γ_k is sufficiently close to a homoclinic orbit, then the above hypothesis is always satisfied. (See Brunovsky and Chow [1]).

If we define the 1-periodic function $h_k(\alpha)$ by

$$\eta_k = \int_0^k \dot{p}_k^2(t)\,dt,$$

(7.4)

$$h_k(\alpha) = \eta_k^{-1} \int_0^k \dot{p}_k(t)F(t-\alpha)\,dt,$$

then we can prove the following result due to Hale and Táboas [1] (for an abstract version, see Hale [9]).

THEOREM 7.1. *Let α_m, α_M be respectively the value of α in $[0, 1)$ for which $h_k(\alpha)$ is a minimum, maximum and suppose $h_k''(\alpha_m) > 0$, $h_k''(\alpha_M) < 0$. Then there are neighborhoods U of Γ_k, V of $(\lambda, \mu) = (0, 0)$ and two curves C_m^k, $C_M^k \subseteq V$ containing $(0, 0)$, tangent respectively to the lines $\lambda = h_k(\alpha_m)\mu$, $\lambda = h_k(\alpha_M)\mu$ at $(0, 0)$, dividing $V\backslash\{0\}$ into two disjoint open sets S_1^k, S_2^k such that equation (7.1) has no subharmonic solution in U for $(\lambda, \mu) \in S_1^k$ and at least $2k$ for (λ, μ) in S_2^k.*

We remark that the hypothesis on h is generic in $F \in C(\mathbf{R}, \mathbf{R})$, $F(t+1) = F(t)$.

In the proof of this theorem, it is shown that $2k$ subharmonic solutions appear as λ/μ decreases through a maximum (increases through a minimum) of $h_k(\alpha)$. Thus, if the function $h_k(\alpha)$ has no other maxima or minima in $[0, 1)$, then the conclusion in Theorem 7.1 can be strengthened to say there are exactly $2k$ subharmonic solutions for $(\lambda, \mu) \in S_2^k$.

Only the idea of the proof will be given. There is a neighborhood U of Γ_k, $\delta > 0$, such that the mapping

$$x = p(\alpha) - a\ddot{p}(\alpha), \qquad y = \dot{p}(\alpha) + a\dot{p}(\alpha)$$

is a one-to-one C' transformation from U onto $[0, k) \times \{|a| < \delta\}$. Thus, it is sufficient to consider only k-periodic solutions of the form

$$x(t) = p(t+\alpha) + z(t+\alpha), \qquad y(t) = \dot{x}(t),$$

$$z(\alpha) = -a\ddot{p}(\alpha), \qquad \dot{z}(\alpha) = a\dot{p}(\alpha)$$

with $|a| < \delta$, $\alpha \in [0, k)$. If this change of variables is performed, then the method of Liapunov-Schmidt can be applied to obtain the usual bifurcation function $B(\alpha, \lambda, \mu)$. This function has the form

$$B(\alpha, \lambda, \mu) = -\lambda\eta_k + \mu\eta_k h_k(\alpha) + O(|\eta| + |\mu|)^2 \quad \text{as } |\lambda|, |\mu| \longrightarrow 0.$$

The remaining argument is a careful analysis of the zeros of this function (see Hale and Táboas [1]).

For a given α, λ, μ for which $B(\alpha, \lambda, \mu) = 0$, one obtains a k-periodic solution which is close to $(p(t+\alpha), \dot{p}(t+\alpha))$. The constant α corresponds to a phase shift and is approximately determined from the relation $h_k(\alpha) = \lambda/\mu$. If $h_k'(\alpha) \neq 0$ for $\alpha \neq \alpha_m$, α_M, then one can associate a unique $\alpha = \alpha^*(\lambda/\mu)$ with λ/μ. As $\lambda, \mu \longrightarrow 0$, this function $\alpha^*(\lambda/\mu)$ will approach a limit if and only if λ/μ approaches a constant. This implies that the initial values of the corresponding k-periodic solution (which are approximately $p(\alpha^*(\lambda/\mu), \dot{p}(\alpha^*(\lambda/\mu)))$ will converge to a point on Γ_k as $\lambda, \mu \longrightarrow 0$ if and only if λ/μ approaches a constant as

$\lambda, \mu \longrightarrow 0$. This points out one of the disadvantages of treating the original two parameter problems in (λ, μ) as a one parameter problem along a ray $(\lambda, m\lambda)$, m fixed, in parameter space.

Using the results of §4, if $h_k(\alpha)$ has a finite number of extremal values in $[0, 1)$, one can show that there are at least $2k$ hyperbolic subharmonics of order k with k being saddles and k being either nodes or foci (see, for example, Chow and Hale [1]).

A specific example of equation (7.1) is

$$\ddot{x} - x + x^2 = -\lambda\dot{x} + \mu F(t). \tag{7.5}$$

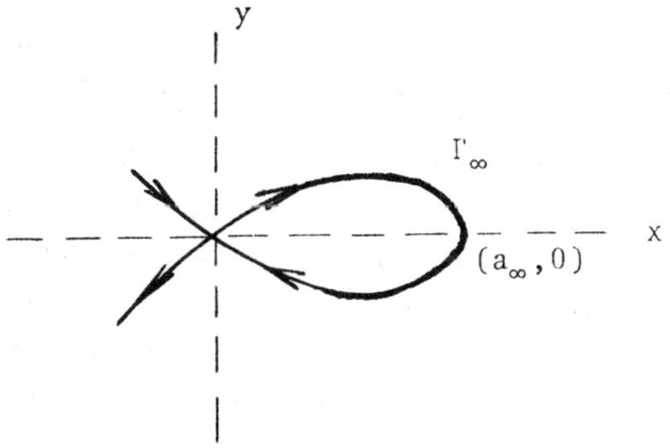

FIGURE 10.

For $(\lambda, \mu) = (0, 0)$, the phase portrait is shown in Figure 10 with any neighborhood of the homoclinic orbit Γ_∞ containing periodic orbits $\Gamma_{\omega(a)}$ with period $\omega(a)$, where $(a, 0)$ is the initial value of the solution at $t = 0$. One can show that $\omega(a) \longrightarrow \infty$ as $a \longrightarrow a_\infty$ monotonically. Thus, there is a sequence $a_k \longrightarrow a_\infty$ such that $\omega(a_k) = k$ for $k \geqslant k_0$ $(= 7)$. The previous results can be applied to each such Γ_k for the subharmonic bifurcation curves C^k_m, C^k_M. This suggests that the behavior of solutions of (7.1) near Γ_∞ could be very complicated. We show that this is actually the case.

Suppose $f(0) = 0$, $f'(0) < 0$ and there is a homoclinic orbit Γ_∞ through $(0, 0)$, $\Gamma_\infty = \{(p_\infty(t), \dot{p}_\infty(t)), t \in \mathbf{R}\} \cup \{(0, 0)\}$, where $p_\infty(t) \longrightarrow 0$ as $t \longrightarrow \pm\infty$ is a solution of (7.2). Since $(0, 0)$ is a saddle point, there are a neighborhood V of $(\lambda, \mu) = (0, 0)$ and a 1-periodic solution $(\phi(t, \lambda, \mu), \dot{\phi}(t, \lambda, \mu))$ of (7.1) for $(\lambda, \mu) \in V$, $\phi(t, 0, 0) = 0$. Furthermore, the trajectory $\gamma_{\lambda,\mu} \subset \mathbf{R} \times \mathbf{R}^2$ of this solution has stable manifold $S_{\lambda,\mu}$ and unstable manifold $U_{\lambda,\mu}$ with 1-periodic cross-sections

$$S_{\lambda,\mu}(t) = \{(x, y): (t, x, y) \in S_{\lambda,\mu}\}, \quad U_{\lambda,\mu}(t) = \{(x, y): (t, x, y) \in U_{\lambda,\mu}\}.$$

Let π be the period one map of equation (7.1). Then $p_{\lambda,\mu} = (\phi(0, \lambda, \mu), \dot{\phi}(0, \lambda, \mu))$ is a fixed point of π and $S_{\lambda,\mu}(0)$, $U_{\lambda,\mu}(0)$ are the stable and unstable manifolds of p.

Problem HB. *Find a neighborhood U of Γ_∞ and a neighborhood V of $(\lambda, \mu) = (0, 0)$ such that, for each $(\lambda, \mu) \in V$, one knows whether or not there is a point $q \in U$ which is transverse homoclinic to $p_{\lambda,\mu}$.*

As noted in §3, the existence of a point which is transverse homoclinic to $p_{\lambda,\mu}$ implies a type of random motion occurs in a neighborhood of Γ_∞. Our primary concern here is to show how the transverse homoclinic point occurs through bifurcation and the connection with the previous problem of subharmonic bifurcation.

Define the 1-periodic function $h_\infty(\alpha)$ by

$$\eta_\infty = \int_{-\infty}^{\infty} \dot{p}_\infty^2(t)\,dt, \qquad h_\infty(\alpha) = \eta_\infty^{-1} \int_{-\infty}^{\infty} \dot{p}_\infty(t) F(t-\alpha)\,dt \tag{7.6}$$

and assume that

$$h''(\alpha_m) > 0, \qquad h''(\alpha_M) < 0 \tag{7.7}$$

where α_m, α_M are the values of α at which h assumes respectively its minimum, maximum. We have the following result of Chow, Hale and Mallet-Paret [2].

THEOREM 7.2. *There are neighborhoods U of Γ_∞, V of $(\lambda, \mu) = (0, 0)$ and two curves C_m^∞, $C_M^\infty \subseteq V$ containing $(0, 0)$ tangent respectively to the lines $\lambda = h_\infty(\alpha_m)\mu$, $\lambda = h_\infty(\alpha_M)\mu$ at $(0, 0)$, dividing $V\backslash\{(0, 0)\}$ into two disjoint open sets S_1^∞, S_2^∞ such that equation (7.1) has no homoclinic points in S_1^∞ and has transverse homoclinic points in S_2^∞.*

In any neighborhood of any point (λ, μ) in the bifurcation set $C_m^\infty \cup C_M^\infty$, there are infinitely many subharmonic bifurcations of the type in Theorem 7.1.

Finally, if $h'_\infty(\alpha)$ has only a finite number of zeros in $[0, 1)$, then the period one map π has infinitely many hyperbolic saddles and nodes (or foci) in U for $(\lambda, \mu) \in S_2^\infty$.

Note the conditions on h_∞ are generic in F.

The proof of the first part of this theorem follows along the same lines as the proof of Theorem 7.1 for subharmonic solutions. One must replace the Fredholm alternative for periodic solutions by a Fredholm alternative for solutions bounded on $\mathbf{R}$. This gives rise to the function $h_\infty(\alpha)$ in (7.6), a function whose importance in this problem was emphasized by Mel'nikov [1]. To prove the part on subharmonic solutions, one shows that condition (7.7) implies the corresponding conditions in Theorem 7.1 are satisfied for each $h_k(\alpha)$ for which $\Gamma_k \subseteq U$. See Chow, Hale and Mallet-Paret [1] for details.

Holmes [1] has also proved the first part of Theorem 7.2 dealing with the existence of homoclinic points. Closely related results for Hamiltonian systems have been obtained by Churchill, Pecelli and Rod [1]. Holmes and Marsden [1] have extended the first part of this theorem to certain types of partial differential equations and have applied the results to the equations for a beam. In this case, technical difficulties primarily arise from the fact that the semigroup generated by the linear part of the unperturbed system has spectrum which can include the complete unit circle. A new phenomenon arises because the spectrum contains one. This can also occur in the finite dimensional case. For example, consider the equations in $\mathbf{R}^3$,

$$\ddot{x} - x + x^2 = -\lambda\dot{x} + \mu F(t) + H(x, y), \quad \dot{y} = -\lambda y + \mu G(t) + L(x, y)$$

where F, G have period 1, H, L vanish together with their first derivatives at $x = 0, y = 0$ and $H(x, 0) = O(|x|^3)$ as $|x| \longrightarrow 0$, $L(x, 0) = 0$. For $\lambda = \mu = 0$, this equation has a homoclinic orbit in the $(x, \dot{x})$-plane. The equilibrium point $x = \dot{x} = 0, y = 0$ is not a saddle since one eigenvalue is zero. To discuss homoclinic orbits for the perturbed equation, one must find first a periodic solution of period one. The difficulties involved are the same as the ones encountered in finding a 1-periodic solution which approaches zero as $\lambda, \mu \longrightarrow 0$ of an equation of the form $\dot{y} = -\lambda y + \mu G(t) + M(y)$ where $M(y) = O(|y|^2)$ as $|y| \longrightarrow 0$. Given the first nonvanishing coefficient of $M(y)$ and assuming that $\int_0^1 G(t)\,dt \neq 0$, this puts restrictions on the parameters λ, μ. For example, if $M(y) = \beta y^k + O(|y|^{k+1})$, $\beta \neq 0$, then $k = 2$ implies μ/λ^3 approaches a constant as $\lambda \longrightarrow 0$. If $k = 3$, then μ/λ^2 approaches a constant as $\lambda \longrightarrow 0$. Once the 1-periodic solution has been obtained, it is not too difficult to adapt the previous ideas to obtain a transverse homoclinic point in some region of the parameter space.

It often occurs that one is interested in studying problems where the orbit Γ_∞ is not a homoclinic orbit, but is the closure of an orbit connecting two distinct equilibrium points. To obtain a generalization of Theorem 7.2 for this case, we give a procedure motivating the appropriate formula for the corresponding function $h_\infty(\alpha)$. Consider the equation (7.5) with $\lambda = 0$. Suppose there is a homoclinic orbit for λ small. It must be close to p_∞. If it has the form $p_\infty + \mu q + o(|\mu|)$ then q satisfies the equation $\ddot{q} + f'(p_\infty)q = F(t)$. Multiply by $\dot{p}_\infty$, integrate from 0 to ∞, and integrate the term $\dot{p}_\infty \ddot{q}$ by parts twice. The result is

$$-\ddot{p}_\infty(0)q(0) = \int_0^\infty F(t)\dot{p}_\infty(t)\,dt.$$

Doing the same thing but integrating from $-\infty$ to zero, one has $\ddot{p}_\infty(0)q(0) = \int_{-\infty}^0 F(t)\dot{p}_\infty(t)\,dt$. Since $\ddot{p}_\infty(0) \neq 0$, this implies $\int_{-\infty}^\infty F(t)\dot{p}_\infty(t)\,dt = 0$, which is the same as $h_\infty(0) = 0$.

Let us turn our attention now to the bifurcation of tori. Suppose $G\colon \mathbf{R}^2 \longrightarrow \mathbf{R}^2$, $F\colon \mathbf{R} \times \mathbf{R}^2 \times \mathbf{R}^k \longrightarrow \mathbf{R}^2$ are C^r-functions, $r \geqslant 2$, $F(t, y, \epsilon)$ periodic in t, $F(t, y, 0) = 0$, and consider the equation

$$\dot{y} = G(y) + F(t, y, \epsilon). \tag{7.8}$$

For $\epsilon = 0$, suppose this equation has an ω-periodic orbit Γ. In a neighborhood of Γ, one can introduce a moving orthonormal coordinate system (ρ, θ) which is ω-periodic in θ. The new equations for (ρ, θ) have the form

$$\dot{\theta} = 1 + \Theta(t, \theta, \rho, \epsilon), \qquad \dot{\rho} = R(t, \theta, \rho, \epsilon) \tag{7.9}$$

where Θ, R are periodic in t, θ, $\Theta(t, \theta, 0, 0) = 0$, $R(t, \theta, \rho, 0) = a\rho + O(|\rho|^2)$ as $|\rho| \longrightarrow 0$. If $a \neq 0$, that is, the orbit Γ of (7.8) for $\epsilon = 0$ is hyperbolic, then the classical results on the theory of integral manifolds (see, for example, Hale [1]) imply the existence of a function $\rho^*(t, \theta, \epsilon)$, $\rho^*(t, \theta, 0) = 0$ defined for all t, ϵ and periodic in both variables such that $(\theta(t), \rho^*(t, \theta(t), \epsilon))$ is a solution of (7.9) for every solution $\theta(t)$ of

$$\dot{\theta} = 1 + \Theta(t, \theta, \rho^*(t, \theta, \epsilon), \epsilon).$$

For equation (7.8), this implies the existence of an invariant cylinder in $\mathbf{R} \times \mathbf{R}^2$ with periodic cross-section near the cylinder $\mathbf{R} \times \Gamma$. Since the cross-section is periodic, the flow on the cylinder is equivalent to the existence of a flow on a torus. If the original orbit Γ is not hyperbolic, one must determine more explicitly how the nonlinear terms influence the flow. We return to this problem later.

Suppose $H: \mathbf{R}^3 \times \mathbf{R}^k \longrightarrow \mathbf{R}^3$ and consider the equation

$$\dot{z} = H(z, \epsilon). \tag{7.10}$$

Suppose for $\epsilon = 0$ this equation has a smooth invariant torus T^2 on which the flow is parallel; that is, either the rotation number is irrational or every orbit is periodic. It is possible to introduce coordinates (ρ, θ, ζ) in a neighborhood of T^2 such that equation (7.10) is equivalent to

$$\dot{\zeta} = 1 + Z(\zeta, \theta, \rho, \epsilon), \qquad \dot{\theta} = 1 + \Theta(\zeta, \theta, \rho, \epsilon), \qquad \dot{\rho} = R(\zeta, \theta, \rho, \epsilon) \tag{7.11}$$

where all functions are periodic in ζ, θ and vanish for $(\rho, \epsilon) = (0, 0)$. Any integral manifold of equation (7.11) which can be represented by a function $\rho^*(\zeta, \theta, \epsilon)$, periodic in ζ, θ with the same period as the vector field, will correspond to an invariant torus of (7.10).

If the original torus T^2 is hyperbolic, then equation (7.10) will have a unique invariant torus in a neighborhood of T^2. We will discuss some aspects of the problem when T^2 is not hyperbolic. Note that equation (7.9) can be considered as a special case of equation (7.11) by putting $t = \zeta$.

To be specific, suppose $\epsilon = (\lambda, \mu) \in \mathbf{R}^2$ and

$$R(\zeta, \theta, \rho, \lambda, \mu) = \rho^2 + \mu f(\theta, \zeta) + \lambda g(\theta, \zeta) + \widetilde{R}(\zeta, \theta, \rho, \lambda, \mu) \tag{7.12}$$

where $\widetilde{R} = O(|\rho|^3 + (|\mu| + |\lambda|)(|\mu| + |\lambda| + |\rho|))$ as $|\lambda|, |\mu|, |\rho| \longrightarrow 0$. For $(\lambda, \mu) = (0, 0)$, equation (7.11) has the invariant torus $\rho = 0$ which is asymptotically stable from one side and unstable from the other. One would suspect that there are no invariant tori for some (λ, μ) near zero and two invariant tori for other (λ, μ); that is, there is a bifurcation of tori similar to the saddle-node type bifurcation for equilibrium points. We will confirm some of these suspicions and also point out that other phenomena can occur.

Let us first consider the one parameter problem with $\lambda = 0$. If

$$f_0 = \lim_{T \to \infty} \frac{1}{T} \int_0^T f(\theta + t, \zeta + t)\, dt \tag{7.13}$$

is independent of (θ, ζ) (this is a nonresonance condition), then the scaling $\rho \longrightarrow \sqrt{|\mu|}\rho$ can be justified. Using the standard method of averaging (see Hale [1], Diliberto [1]), there is a transformation of variables which takes (7.11) into an equivalent equation

$$\dot{\zeta} = 1 + O(|\mu|^{1/2}), \qquad \dot{\theta} = 1 + O(|\mu|^{1/2}),$$

$$\dot{\rho} = |\mu|^{1/2}(\rho^2 + f_0 \operatorname{sgn} \mu) + o(|\mu|^{1/2})$$

as $|\mu| \longrightarrow 0$. Integral manifold theory implies there are no invariant tori if $f_0 \operatorname{sgn} \mu > 0$ and two hyperbolic invariant tori if $f_0 \operatorname{sgn} \mu < 0$.

If $g_0 = \lim_{T\to\infty} T^{-1}\int_0^T g(\theta + t, \zeta + t)dt$ is independent of θ, ζ and $\lambda = \gamma\hat{\mu}$, then one can apply the same result to the complete equation (7.11) with the only change being that $f_0 \operatorname{sgn} \mu$ is replaced by $(f_0 + \gamma g_0)\operatorname{sgn} \mu$. If this quantity is > 0, there are no invariant tori and if this quantity is < 0, there are two invariant tori.

From the above discussion, we have obtained the following result. *If $f_0 \neq 0$, $g_0 \neq 0$, then, for any γ for which $f_0 + \mu g_0 \neq 0$, there is a $\mu_0 = \mu_0(\gamma, f_0, g_0)$ such that on the line segment $L_\gamma = \{(\lambda, \mu)\colon \lambda = \gamma\mu, 0 \leqslant |\mu| \leqslant \mu_0\}$, there is no invariant torus of* (7.11) *in a neighborhood of zero if $(f_0 + \gamma g_0)\operatorname{sgn} \mu > 0$ and there are two hyperbolic tori if* $(f_0 + \gamma g_0)\operatorname{sgn} \mu > 0$.

This result implies a complete solution of the problem of bifurcation of tori except in a sector S_{γ_0} containing the line segment L_{γ_0}, $\gamma_0 = -f_0/g_0$ (see Figure 11). This sector must be tangent to L_{γ_0} at (0, 0). If one assumes the vector field is C^∞ with the coefficients in the Taylor series in ρ being trigonometric polynomials and the ratio of the periods in θ, ζ are irrational, then it is possible to show that the excluded sector S_{γ_0} is tangent to L_{γ_0} at (0, 0) to infinite order.

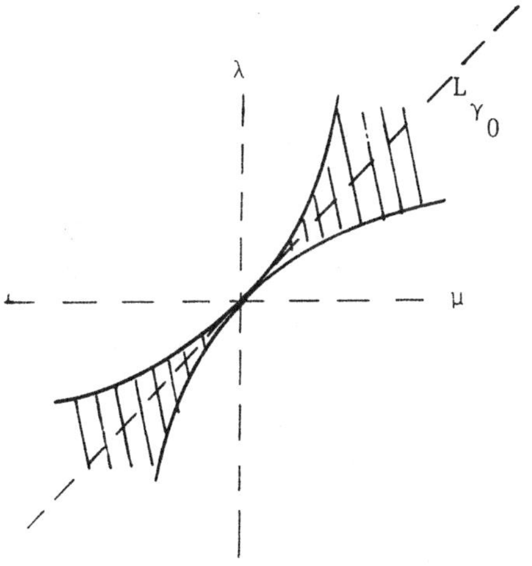

FIGURE 11.

What happens in S_{γ_0} is very complicated. In fact, when the two tori come too close together, then generically the rotation number is rational and hyperbolic periodic orbits appear. The two invariant sets then can meet at these hyperbolic periodic orbits and then change topological structure. An illustrative example of the cross-sections of these sets is shown in Figure 12 as the generic bifurcation takes place. The interior set is exponentially stable in Figure 12a and the exterior one is unstable exponentially. The points corresponding to the periodic orbits are hyperbolic–all being saddle points except for the nodes *A, B, C, D.* In Figure 12b, the saddles remain and thepoints *A, B* are saddle-nodes. In Figure 12c, the invariant sets that were similar to tori have disappeared leaving only hyperbolic orbits.

The complete structure of the solutions in S_{γ_0} is not known. The author is indebted to Brunovksy and Chow for conversations about this problem.

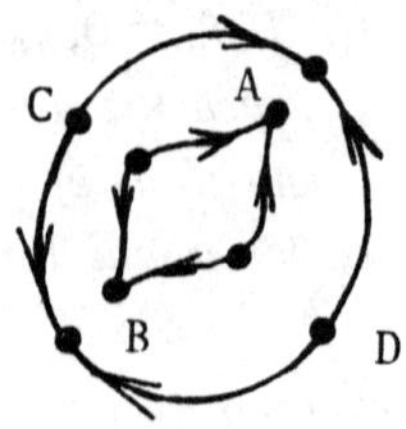

FIGURE 12a.

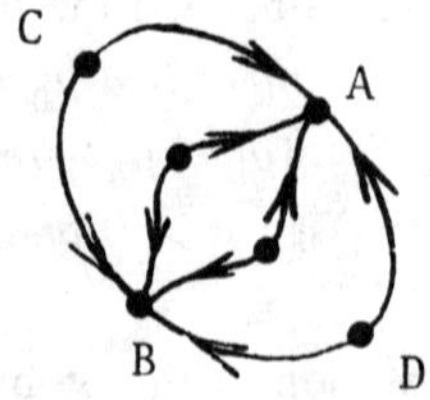

FIGURE 12b.

FIGURE 12c.

8. Higher order bifurcation near equilibrium

In §6, the vector fields in the plane corresponding to bifurcation points of order one were completely characterized. Higher order bifurcation in any dimension were discussed in §§4, 5 for the case of a generalized saddle-node or focus. In this section, we give two illustrations of a bifurcation point of order two near an equilibrium point for which the complete analysis requires concepts which are global in nature–in particular, involve knowledge of homoclinic orbits and invariant tori. One vector field is two dimensional for which the matrix of the linear approximation near zero has both roots zero but with nonsimple elementary divisors and is based on Howard and Kopell [1]. The second illustration is a three dimensional vector field with the matrix of the linear approximation near zero having two purely imaginary and one zero root. The relevant literature for this latter example is Langford [1], Guckenheimer [3], Chow and Hale [1].

Consider the equation

(8.1) $$\dot{x} = y, \qquad \dot{y} = \epsilon_1 x + \epsilon_2 y + \alpha x^2 + \beta xy$$

where $\alpha < 0, \beta > 0$ are fixed constants and $\epsilon = (\epsilon_1, \epsilon_2)$ is real, varying in a neighborhood of zero.

Knowing the behavior of (8.1) will also determine the behavior for the case $\alpha \neq 0$, $\beta \neq 0$ since the change of variables $t \longrightarrow -t$, $\epsilon_2 \longrightarrow -\epsilon_2$, $x \longrightarrow \pm x$, $y \longrightarrow \mp y$ will yield the above case. Also, the form (8.1) is chosen for simplicity and the qualitative structure will be unchanged if perturbations (X, Y) in the vector field are made which satisfy $X = O((|x| + |y|)^2 + |\epsilon|(|x| + |y|))$, $Y = O((|x| + |y|)^2(1 + |\epsilon|))$.

The objective is to discuss the qualitative behavior of the solutions of (8.1) in a neighborhood of $(x, y) = (0, 0)$ for ϵ varying in a neighborhood of zero.

First, we consider the case $\epsilon_1 > 0$. The introduction of the scaled variables

(8.2) $$\begin{gathered} \epsilon_1 = \delta^2, \qquad \epsilon_2 = \mu\delta^2, \qquad \delta > 0, \\ t \mapsto \delta^{-1} t, \qquad x \mapsto \delta^2 |\alpha|^{-1} x, \qquad y \mapsto \delta^3 |\alpha|^{-1} y \end{gathered}$$

leads to the new equations

(8.3) $$\dot{x} = y, \qquad \dot{y} = x - x^2 + \mu\delta y + \delta\gamma xy$$

where $\gamma = \beta|\alpha|^{-1}$.

For $\delta = 0$, equation (8.3) is conservative with the first integral $V(x, y) = y^2/2 - x^2/2 + x^3/3$. The equilibrium point $(0, 0)$ is a saddle point with a homoclinic orbit Γ_∞ through it while the equilibrium point $(1, 0)$ is a center. The first step is to analyze the periodic orbits of equation (8.3).

LEMMA 8.1. *Every periodic orbit of equation* (8.3) *must intersect the segment* $(0, 1) \times \{0\}$ *in the* (x, y)*-plane. There are a continuous positive function* $\delta_0(b)$, $b \in (0, 1)$, *and a continuously differentiable function* $\mu^*(b, \delta)$, $b \in (0, 1)$, $|\delta| < \delta_0(b)$ *such that equation* (8.3) *has a periodic orbit if and only if* $\mu = \mu^*(b, \delta)$. *Furthermore*, $d\mu^*(b, 0)/db < 0$, $\mu^*(b, 0) \longrightarrow -\gamma$ *as* $b \longrightarrow 1$, $\mu^*(b, 0) \longrightarrow -6\gamma/7$ *as* $b \longrightarrow 0$. *Finally, if* $\mu = \mu^*(b, \delta)$ *for a fixed* $b \in (0, 1)$ *and* $|\delta| < \delta_0(b)$, *then the periodic orbit through* $(b, 0)$ *is the only one corresponding to this* μ, δ.

Only a sketch of the proof is given (see Chow and Hale [1] for details based on Carr [1]). The first statement is obvious since the index of a periodic orbit is one. If $b \in (0, 1)$ is on a periodic orbit Γ, then $\int_\Gamma \dot{V} dt = 0$. Using the fact that the orbit Γ must be symmetric with respect to the x-axis, this implies that

$$\mu \int_b^{c(b)} y^2\, dx + \gamma \int_b^{c(b)} xy^2\, dx = 0$$

where $(c(b), 0) \in \Gamma$, $c(b) > 1$. This defines $\mu^*(b, \delta)$. The other assertions require a number of computations which will not be given.

Let $\Gamma_\infty = \{(p_\infty(t), \dot{p}_\infty(t)), t \in \mathbf{R}\} \cup \{(0, 0)\}$ where p_∞ is a solution of (8.3) for $\epsilon = 0$, $p_\infty(t) \longrightarrow 0$ as $t \longrightarrow \pm\infty$. Using the type of analysis in §7, one can find a unique curve in the (μ, δ)-plane for δ small along which equation (8.3) has a homoclinic orbit. This gives a curve C_∞ in the $\epsilon = (\epsilon_1, \epsilon_2)$-plane defined for $0 \leqslant |\epsilon| \leqslant \epsilon_0$ parametrically by a function $\epsilon_2 = c(\epsilon_1^{1/2})\epsilon_1$ where $c(0) = -\gamma\nu$, $\nu = \int_{-\infty}^{\infty} q_\infty \dot{q}_\infty^2 / \int_{-\infty}^{\infty} \dot{q}^2$. On C_∞, the homoclinic orbit is asymptotically stable from §6. Using the formulas which define the homoclinic orbit from Chow, Hale and Mallet-Paret [2], one shows that there is a unique periodic orbit to the left of C_∞. To the right of C_∞, every solution of (8.3) leaves a neighborhood of Γ_∞.

To the left of C_∞ and close to C_∞ where the periodic orbit exists, this orbit passes through $(b, 0)$ where b is close to zero. Lemma 2.1 implies that this orbit continues to exist and is the unique periodic orbit on the ray $\epsilon_2 = \mu^*(b, \delta)\epsilon_1$. Thus, as long as b remains in an interval $(0, 1 - d]$, $d \in (0, 1)$, we have the existence on a uniform δ-interval $0 \leqslant |\delta| \leqslant \delta_0(d)$. To obtain a uniform δ-interval for $b \in (0, 1)$, we analyze carefully the neighborhood of the equilibrium point $(1, 0)$. The matrix of the linear variational equation near the equilibrium point $(1, 0)$ has complex eigenvalues with real parts $\delta(\mu + \gamma)$ and are $\pm i$ at $\delta = 0$. The bifurcation function $G(a, \delta, \mu)$ for periodic orbits obtained as in §5 is an analytic function of (a, δ, μ) near $(a, 0, -\gamma)$. Also, $G(a, 0, \mu) = 0$ for all (a, μ) since equation (8.3) for $\delta = 0$ has a first integral (Corollary 5.4). Since $G(0, \delta, \mu) = 0$ for all δ, μ, this implies that the bifurcation equation $G(a, \delta, \mu) = 0$ can be replaced by the equivalent equation $H(a, \delta, \mu) = 0$, where $H(a, \delta, \mu) = G(a, \delta, \mu)/a\delta$. Also, $H(0, 0, \mu) = \mu + \delta +$ (terms vanishing for $(a, \delta) = (0, 0)$). The Implicit Function Theorem implies there is a unique function $\mu^*(a, \delta)$ such that $H(a, \delta, \mu^*(a, \delta)) = 0$ for $|a| < a_0$, $|\delta| < \delta_0$. This gives the additional uniformity in δ for b near 1.

Combining all of this information, the half-plane $\epsilon_1 > 0$ can be divided into four sectors as shown in Figure 13a for which the corresponding flows are given in Figures 13b, c, d.

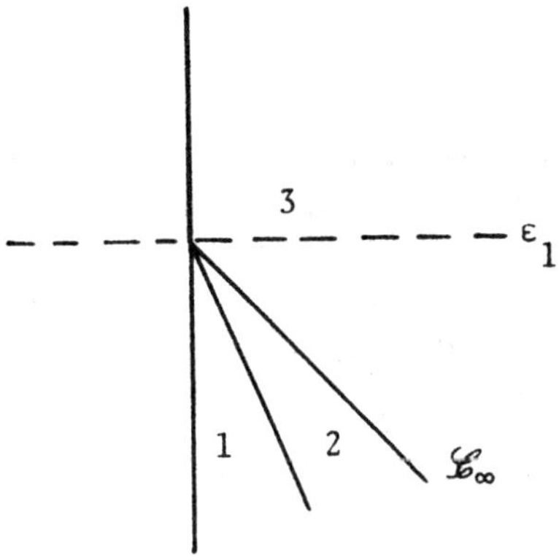

FIGURE 13a.

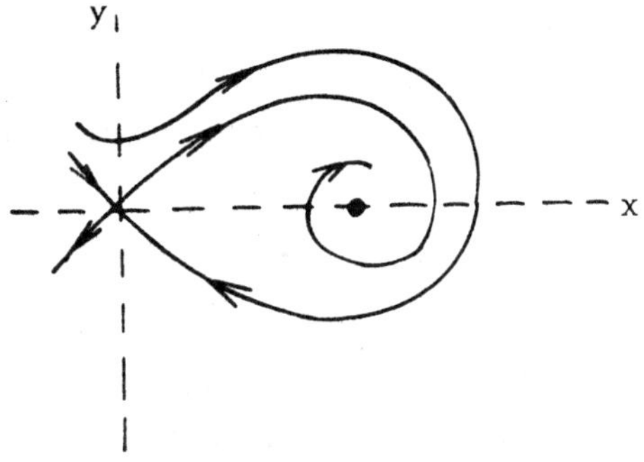

FIGURE 13b. Sector 1.

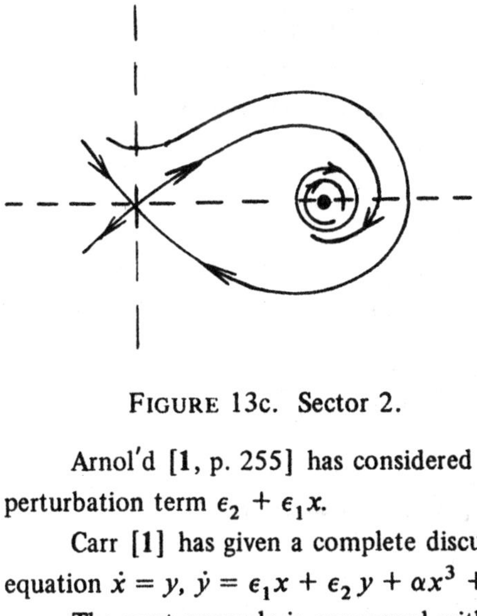

FIGURE 13c. Sector 2.

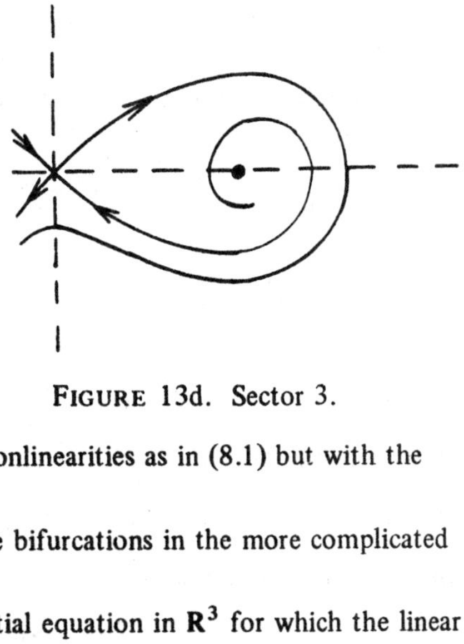

FIGURE 13d. Sector 3.

Arnol'd [1, p. 255] has considered the same nonlinearities as in (8.1) but with the perturbation term $\epsilon_2 + \epsilon_1 x$.

Carr [1] has given a complete discussion of the bifurcations in the more complicated equation $\dot{x} = y$, $\dot{y} = \epsilon_1 x + \epsilon_2 y + \alpha x^3 + \beta x^2 y$.

The next example is concerned with a differential equation in $\mathbf{R}^3$ for which the linear variational equation near the equilibrium point zero has two purely imaginary eigenvalues and one zero eigenvalue. To simplify the situation, it will be assumed also that a certain type of symmetry prevails. More specifically, consider the equation

$$\dot{x} = A(\lambda)x + f(x, y), \qquad \dot{y} = \beta y + g(x, y) \tag{8.4}$$

where λ, β are small real parameters, $x \in \mathbf{R}^2$, $y \in \mathbf{R}$, f, g are C^4-functions,

$$A(\lambda) = \begin{pmatrix} \lambda & 1 \\ -1 & \lambda \end{pmatrix}, \tag{8.5}$$

$$f(x, y) = O(|x|(|x| + |y|)), \qquad g(x, y) = O(|x| + |y|)^2$$

as $|x|, |y| \longrightarrow 0$. The hypothesis on f implies that the symmetry condition

$$f(0, y) = 0 \tag{8.6}$$

is satisfied.

Since $f(0, y) = 0$, it is legitimate to introduce polar coordinates for $x = (x_1, x_2)$ as $x_1 = \rho \cos \theta$, $x_2 = -\rho \sin \theta$. If this is done and t is replaced by θ, one obtains the equations

$$\dot{\rho} = \lambda\rho + R(\theta, \rho, y, \alpha, \beta), \qquad \dot{y} = \beta y + Y(\theta, \rho, y, \alpha, \beta) \tag{8.7}$$

where $R = O(|\rho|(|\rho| + |y|))$, $Y = O((|\rho| + |y|)^2)$ as $\rho, y \longrightarrow 0$.

The problem is to determine the behavior of the solutions of (8.7) in a neighborhood of $(\rho, y) = (0, 0)$ for (λ, β) in a neighborhood of $(0, 0)$. To discuss (8.7), we suppose that an application of the theory of normal forms to (8.7) yields the system

$$\dot{\rho} = \rho(\lambda + ay + b\rho^2) + O(\rho(|\rho| + |y|)^3), \quad \dot{y} = \beta y + cy^2 + d\rho^2 + O((|\rho| + |y|)^4) \tag{8.8}$$

as $\rho, y \longrightarrow 0$. The generic situation is to have a, b, c, d nonzero constants.

To simplify the computations and also to consider the most interesting case, we suppose $a = 2$, $b = 1$, $c = d = -1$, so that the truncated form of (8.8) using only the lowest order terms is

$$\dot{\rho} = \rho(\lambda + 2y + \rho^2), \qquad \dot{y} = \beta y - y^2 - \rho^2. \tag{8.9}$$

The first objective is to discuss the behavior of the solutions of (8.9) as a function of α, β. It is convenient to introduce the change of variables $y \mapsto \beta/2 + y$, $\lambda + \beta \mapsto \alpha$, to obtain the more symmetric form

$$\dot{\rho} = \rho(\alpha + 2y + \rho^2), \qquad \dot{y} = \beta^2/4 - y^2 - \rho^2. \tag{8.10}$$

If we perform the scalings

$$\rho \mapsto \epsilon\rho, \quad y \mapsto \epsilon y, \quad \beta \mapsto \epsilon, \quad \alpha \mapsto \epsilon\alpha, \quad t \mapsto \epsilon^{-1}t \tag{8.11}$$

the new equations become

$$\dot{\rho} = 2\rho y + \alpha\rho + \epsilon\rho^3, \qquad \dot{y} = 1/4 - y^2 - \rho^2. \tag{8.12}$$

Equations (8.12) must be discussed for all $\rho \geqslant 0$, $y \in \mathbf{R}$, $\alpha \in \mathbf{R}$ and ϵ in a neighborhood of $\epsilon = 0$.

For $\alpha = 0$, $\epsilon = 0$, the function

$$V(\rho, y) = \rho/4 - \rho y^2 - \rho^3/3 \tag{8.13}$$

is a first integral of (8.12). Differentiating this function along the solutions of (8.12), it is not difficult to see that no periodic orbits can exist unless $\alpha = 0$ when $\epsilon = 0$. This remark makes much of the discussion of (8.12) very simple since it is only necessary to determine the structure of the equilibrium points away from the critical value $\alpha = 0$, $\epsilon = 0$.

The equilibrium points of (8.12) are $\rho = 0$, $y = \pm 1/2$, for all α, ϵ and the point $y = -\alpha/2$, $\rho^2 = (1 - \alpha^2)/4$ for $\epsilon = 0$, $\alpha^2 \leqslant 1$. For $\alpha^2 \neq 1$ the points $\rho = 0$, $y = \pm 1/2$ are hyperbolic saddles or nodes. Further analysis of the other equilibrium point for $\alpha^2 < 1$ shows that the original (λ, β)-plane, $\lambda = \beta\alpha - \beta$, $\beta = \epsilon$, can be divided into sectors as shown in Figure 14a with the corresponding phase protrait for $\beta > 0$ given in Figure 15. The situation is similar for $\beta > 0$. There is a saddle-node bifurcation on each of the solid curves in Figure 14a. The analysis in a neighborhood of the dotted line $\lambda = -\beta$ requires further study and will now be given.

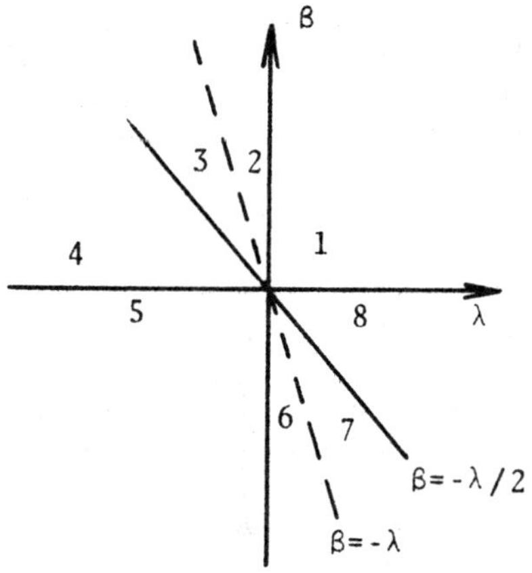

FIGURE 14a.

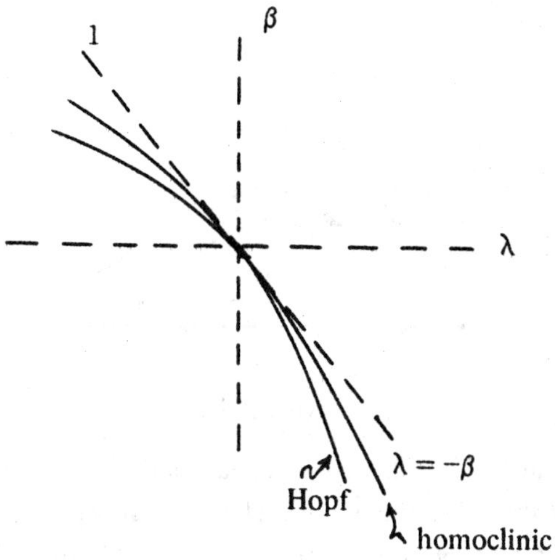

FIGURE 14b.

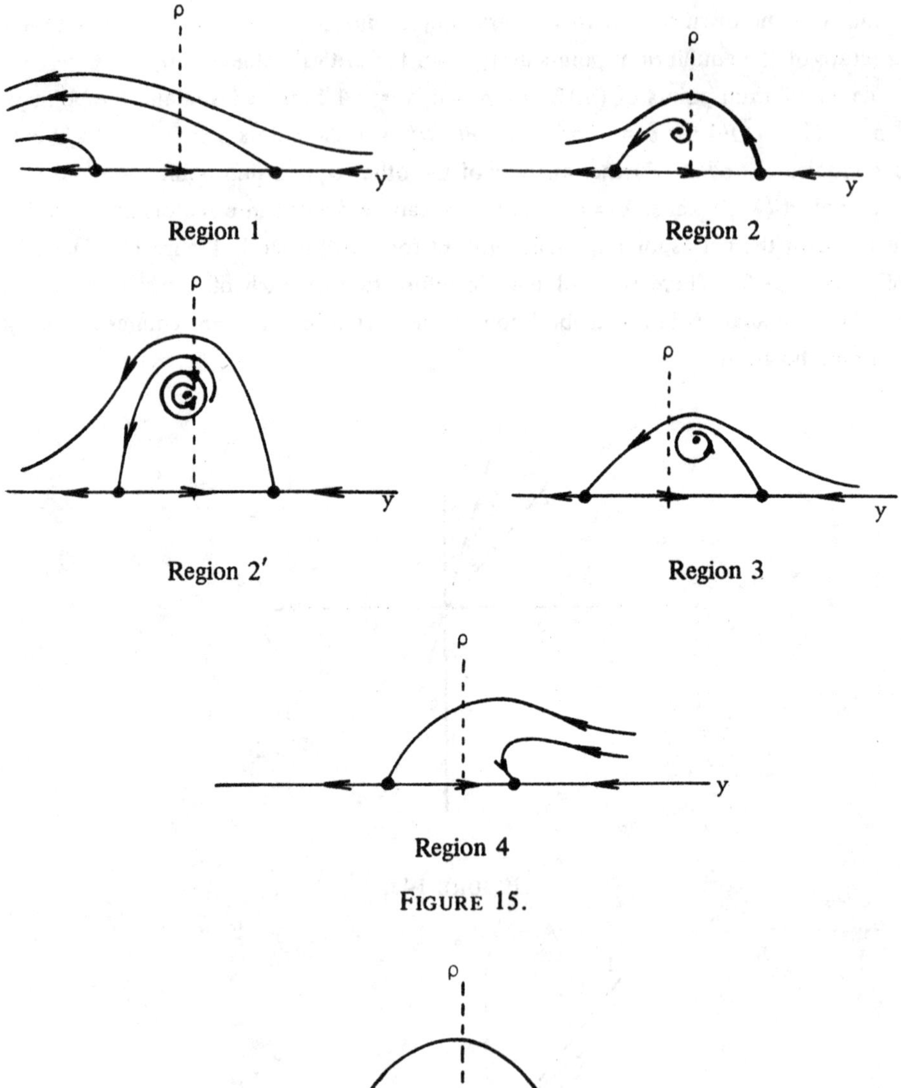

FIGURE 15.

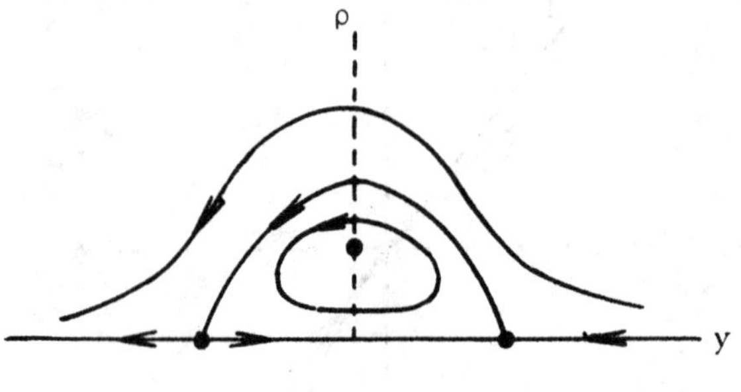

FIGURE 16.

For $\alpha = 0$, $\epsilon = 0$ in (8.12), the phase portrait is determined by the first integral $V(\rho, y)$ in (8.13) and is shown in Figure 16. There is a heteroclinic orbit connecting the two saddle points. Let $(\rho_0(t), y_0(t))$, $\rho_0(t) > 0$, be the solution of (8.12) for $\alpha = \epsilon = 0$ such that $\rho_0(t) \longrightarrow 0$ as $t \longrightarrow \pm\infty$, $y_0(t) \longrightarrow -\frac{1}{2}$ as $t \longrightarrow -\infty$, $y_0(t) \longrightarrow \frac{1}{2}$ as $t \longrightarrow \infty$. Using the remarks after Theorem 7.2, one can obtain a function $G(\alpha, \epsilon)$ for $|\alpha|$, $|\epsilon| < \delta$ such that equation (8.12) has a heteroclinic orbit if and only if $G(\alpha, \epsilon) = 0$. Furthermore, the function G satisfies

$$G(\alpha, \epsilon) = \alpha \int_{-\infty}^{\infty} \rho_0^2(t)\,dt + \epsilon \int_{-\infty}^{\infty} \rho_0^4(t)\,dt + O((|\alpha| + |\epsilon|)^2)$$

as $\alpha, \epsilon \longrightarrow 0$. The Implicit Function Theorem implies there is a unique solution $\alpha = \alpha^*(\epsilon)$ of this equation for ϵ sufficiently small, $\alpha^*(0) = 0$ and

$$-\delta \overset{\text{def}}{=} \frac{d\alpha^*(0)}{d\epsilon} = - \int_{-\infty}^{\infty} \rho_0^4(t)\,dt \Big/ \int_{-\infty}^{\infty} \rho_0^2(t)\,dt = -\frac{1}{2}.$$

This gives a curve $\lambda = \lambda^*(\beta)$ in the original parameter space (λ, β) where there is a heteroclinic orbit for equation (8.10). This curve is given approximately by $\lambda^*(\beta) = -\beta - \delta\beta^2$.

Near the line $\lambda = -\beta$ (or $\alpha = 0$), there is also a Hopf bifurcation. Near $\alpha = 0$, $\epsilon = 0$, equation (8.12) has an equilibrium solution given approximately by $y_0 = -(\alpha + \epsilon/4)/2$, $\rho_0 = 1/2 + y_0$. Analyzing the stability properties of this solution, we see that it has eigenvalues on the imaginary axis along a curve given approximately by $\alpha = -3\epsilon/4$ or $\lambda = -\beta - 3\beta^2/4$. One can actually show there is a Hopf bifurcation along this curve. Also, one can show there is a unique periodic orbit between this curve and the one defining the homoclinic orbit. This implies the complete bifurcation diagram near the line $\lambda = -\beta$ is the one shown in Figure 14b. The flow in region 2′ is shown in Figure 15. The flow on the homoclinic curve is shown in Figure 16. For the details of the above computations see Chow and Hale [1].

It remains to relate the solutions of the approximate equations (8.12) to the complete equations (8.8). Since the perturbation terms are periodic in t, it is not difficult to show that the equilibrium points of (8.12) become periodic solutions of (8.8) and periodic orbits of (8.12) become invariant tori for (8.8). Around the homoclinic orbit, one expects a behavior similar to the one discussed in §7. The precise behavior, of course, depends upon the higher order terms in (8.8) and the method of analysis will be similar to the one in §7.

Equations (8.8) are the generic situation for two purely imaginary and one zero eigenvalue. If further symmetries occur in the problem, there may be no second order terms in the normal form for the vector field. In this case, the simplest case for the approximate equations are

$$\dot{\rho} = \rho(\lambda - ar^2 - by^2), \qquad \dot{y} = y(\beta + cr^2 + dy^2)$$

with a, b, c, d fixed nonzero constants and λ, β small bifurcation parameters. In the case of a fourth order equation with two purely imaginary roots, the same equations occur in a natural way coupled with two angle variables. The complete bifurcation diagram for this equation cannot be obtained without the addition of terms of order five. It is much more difficult and the reader is referred to Holmes [2], Guckenheimer [4], Chow and Hale [1].

9. A framework for infinite dimensions

For infinite dimensional systems, the generic theory analogous to the one discussed in §3 is in its infancy. In this section, we outline an approach to the development of such a theory for a special class of semigroups of transformations. This class is general enough to include some types of parabolic and hyperbolic partial differential equations as well as retarded functional differential equations and some neutral functional differential equations.

Let X, Y, Z be Banach spaces and let $\mathcal{X}^r = C^r(Y, Z)$, $r \geqslant 1$, be the set of functions from Y to Z which are bounded and uniformly continuous together with their derivatives up through order r. We impose the usual topology on $\mathcal{X}^r$. For each $f \in \mathcal{X}^r$, let $T_f(t)\colon X \longrightarrow X$, $t \geqslant 0$ be a strongly continuous semigroup of transformations on X. For each $x \in X$, we suppose $T_f(t)x$ is defined for each $t \geqslant 0$ and is C^r in x.

In applications, one often is interested in open subsets of X, Y, but this presents mainly notational difficulties and therefore will not be discussed.

We say that a point x_0 has a *backward extension* (relative to $T_f(t)$) if there is a function $\phi\colon (-\infty, 0] \longrightarrow X$ such that $\phi(0) = x_0$ and $T_f(t)\phi(\tau) = \phi(t + \tau)$ for all $0 \leqslant t \leqslant -\tau$, $\tau \in (-\infty, 0]$. If x_0 has a backward extension ϕ, we define $T_f(t)x_0$, $t \leqslant 0$, by the relation $T_f(t)x_0 = \phi(t)$ and say that $T_f(t)x_0$ is *defined for* $t \leqslant 0$.

The *positive orbit* $O^+(x_0)$ *through* x_0 is $O^+(x_0) = \bigcup_{t \geqslant 0} T_f(t)x_0$. The *negative orbit* $O^-(x_0)$ *through* x_0 is $O^-(x_0) = \bigcup_{t \leqslant 0} T_f(t)x_0$ if $T_f(t)x_0$ is defined for $t \leqslant 0$. The *ω-limit set* $\omega(x_0)$ *of* x_0 *and α-limit set* $\alpha(x_0)$ *of* x_0 are defined as

$$\omega(x_0) = \bigcap_{\tau \geqslant 0} \operatorname{cl} \bigcup_{t \geqslant \tau} T_f(t)x_0, \quad \alpha(x_0) = \bigcap_{\tau \leqslant 0} \operatorname{Cl} \bigcup_{t \leqslant \tau} T_f(t)x_0.$$

A set $M \subset X$ is *invariant* (*relative* to $T_f(t)$) if, for any $x_0 \in M$, $T_f(t)x_0$ is defined for $t \leqslant 0$ and $T_f(t)x_0 \in M$ for all $t \in (-\infty, \infty)$. An *equilibrium point* of $T_f(t)$ is an invariant set consisting of a single point; that is, an $x_0 \in X$ such that $T_f(t)x_0 = x_0$ for all $t \in \mathbf{R}$. A *periodic orbit* γ of $T_f(t)$ is an invariant set which is a closed curve; that is, there are an $x_0 \in X$ and an $\omega > 0$ such that $T_f(t)x_0 = T_f(t + \omega)x_0$ for $t \in \mathbf{R}$, ω is the least period, and $\gamma = \bigcup_t T_f(t)x_0$.

The following result is easy to prove (see, for example, Hale [5]).

PROPOSITION 9.1. *If $O^+(x_0)$ is precompact, then $\omega(x_0)$ is a nonempty compact connected invariant set and* $\operatorname{dist}(T_f(t)x_0, \omega(x_0)) \longrightarrow 0$ *as $t \longrightarrow \infty$.*

Let

$$A_f = \{x \in X\colon T_f(t)x \text{ is defined and bounded for } t \leqslant 0\}. \tag{9.1}$$

This set contains all equilibrium points, periodic orbits and the ω-limit sets of any compact orbit. The set A_f is invariant. For any bounded open set $N \subset X$, let

$$A_f(N) = \{x \in X\colon T_f(t)x \text{ is defined and belongs to } N \text{ for } t \leqslant 0\}. \tag{9.2}$$

PROPOSITION 9.2. *If A_f is compact, it is maximal compact invariant and attracts points of X. If, in addition, $T_f(t)$ is one-to-one on A_f, then $T_f(t)$ is a continuous group on A_f.*

The proof is elementary.

DEFINITION 9.3. For $f, g \in \mathcal{X}^r$, we say *f is equivalent to g*, $f \sim g$, if there is a homeomorphism $h\colon A_f \longrightarrow A_g$ such that h maps the orbits of $T_f(t)$ on A_f onto the orbits of $T_g(t)$ on A_g preserving the sense of direction in time. An $f \in \mathcal{X}^r$ is *structurally stable* if there is a neighborhood V of f such that $f \sim g$ for every $g \in V$. An $f \in \mathcal{X}^r$ is a *bifurcation point* if it is not structurally stable.

DEFINITION 9.4. For $f, g \in \mathcal{X}^r$, we say f is *locally equivalent* to g in a neighborhood of x_0 if there are neighborhoods N, M of x_0 and a homeomorphism $h\colon A_f(N) \longrightarrow A_g(M)$ such that h takes the orbits of $T_f(t)$ on $A_f(N)$ onto the orbits of $T_g(t)$ on $A_g(M)$ preserving the sense of direction in time. An $f \in \mathcal{X}^r$ is *locally structurally stable* if there is a neighborhood V of f such that f is locally equivalent to g at x_0 for every $g \in V$.

It is important to notice that the flows defined by $T_f(t)$, $T_g(t)$ are compared only on the invariant sets A_f, A_g. However, in order for there to be a homeomorphism between A_f and A_g for each g in a neighborhood of f, the set A_f must have some strong type of stability as an invariant set of $T_f(t)$. We shall see several illustrations of this remark in infinite dimensional problems, but the idea is easily understood by comparing this definition with the one given in §3 for ordinary differential equations in $\mathbf{R}^n$.

Consider the equation

$$\dot{x} = Bx + f(x) \tag{9.3}$$

where $x \in \mathbf{R}^n$, $f(0) = 0$, $\partial f(0)/\partial x = 0$. If W_f^u is the unstable manifold of zero (it could contain only the point zero), suppose there is a neighborhood N of $x = 0$ such that if $x_0 \in N \backslash W_f^u$, then there is a $\tau < 0$ such that $T_f(\tau)x_0 \in \partial N$. It follows that $A_f(N) = W_f^u \cap N$. If f is locally structurally stable at zero according to Definition 9.4, then one can show that the equilibrium point $x = 0$ of (9.3) must be hyperbolic; that is, Re $\lambda B \neq 0$. Also, there is a neighborhood V of f such that, for each $g \in V$, there are a neighborhood M of zero, a unique equilibrium point x_g of $g \in M$ and $A_g(M) = W^u(x_g) \cap M$. In Definition 9.4, the trajectories in a full neighborhood of zero are not compared, and they are not even compared on the stable manifold. It frequently happens in applications in infinite dimensions that the stable manifold is infinite dimensional and the unstable manifold is finite dimensional. Comparison of trajectories on the infinite dimensional parts seems to be impossible. Thus, the definition of equivalence is chosen as above.

To define a large class of semigroups $T_f(t)$ which will guarantee that A_f is compact, maximal invariant and satisfies some stability properties, we need some additional notation.

Let $\alpha(B)$ be the *Kuratowskii measure of noncompactness* of a bounded set B in a Banach space X; that is, $\alpha(B) = \inf\{d > 0\colon B$ has a finite cover with each element of diameter $< d\}$. If $T\colon X \longrightarrow X$ is continuous, we say T is a *conditional α-contraction* if there is a constant $k \in [0, 1)$ such that $\alpha(TB) \leqslant k\alpha(B)$ for all bounded sets B in X for which TB is bounded. The map is an *α-contraction* if it is a conditional α-contraction and takes bounded sets into bounded sets. For general properties of the measure of noncompactness and α-contractions, see Sadovskii [1] or Martin [1]. A map T is *conditionally completely continuous,* if, for any bounded set B for which TB is bounded, the closure of TB is compact. If $T(t)$, $t \geqslant 0$, is a family of maps, we say it is a conditional α-contraction [or conditionally completely continuous] if for each t and each bounded set B for which $\{T(\tau)B,\ 0 \leqslant \tau \leqslant t\}$ is bounded, $\alpha(T(t)B) \leqslant k(t)\alpha(B)$ for some $k(t) \in [0, 1)$ [or the closure of $T(t)B$ is compact].

If $r_e(T)$ denotes the radius of the essential spectrum of a linear operator T and $r_e(T) < 1$, then there is an equivalent norm in X such that $T = S + U$ where S, U are linear operators with $|S| < 1$ and U compact (see Leggett [1], [2], Massatt [3]). Thus, there is an equivalent norm in X such that T is an α-contraction.

The semigroup $T_f(t)$ is a conditional α-contraction if it is a conditional α-contraction for each $t > 0$. If $T_f(t)$ is linear with $r_e(T_f(t)) \leqslant \exp(-\beta t)$ for some $\beta > 0$, then there is an equivalent norm in X such that $T_f(t)$ is an α-contraction for each $t > 0$.

PROPOSITION 9.5. *If $T_f(t)$ is a conditional α-contraction, then each bounded orbit is precompact.*

The proof is trivial since $O^+(x_0)$ is invariant under $T_f(t)$ for each $t > 0$.

A set $K \subset X$ is said to *attract a set $M \subset X$ (relative to $T_f(t)$)* if $\operatorname{dist}(T_f(t)M, K) \longrightarrow 0$ as $t \longrightarrow \infty$. A set K in X is *stable* (relative to $T_f(t)$) if, for any neighborhood U of K, there is a neighborhood V of K such that $T_f(t)V \subset U$, $t \geqslant 0$. A set K in X is *asymptotically stable* (*uniformly asymptotically stable*) if it is stable and there is a neighborhood W of K such that K attracts points of W (K attracts W).

The semigroup $T_f(t)$ is *point* (*local*) (*compact*) *dissipative* if there is a bounded set B in X such that B attracts each point (some neighborhood of each point) (each compact set) of X.

PROPOSITION 9.6. *If $T_f(t)$ is a conditional α-contraction and compact dissipative, then A_f is a maximal compact invariant set and A_f is uniformly asymptotically stable. If $T_f(t)$ is a conditional α-contraction, point dissipative and there is some neighborhood O_x of each $x \in X$ such that $\bigcup_{t \geqslant 0} T_f(t)O_x$ is bounded, then the same conclusion holds. If $T_f(t)$ is point dissipative and conditionally completely continuous, then the same conclusion holds. If, in addition, the orbits of bounded sets are bounded, then A_f attracts bounded sets of X.*

For a proof of these results, see Cooperman [1], Hale [6], Massatt [1].

In many applications, the semigroup $T_f(t)$ is not a general α-contraction but satisfies the condition $T_f(t) = S_f(t) + U_f(t)$ where $S_f(t)$ is a linear contraction for each $t > 0$ and the mapping $U_f(t)$ is conditionally completely continuous. Also, the family of mappings $T_f(t)$ is a semigroup on two Banach spaces X_1, X_2 with X_1 compactly imbedded in X_2. Suppose also that, for any set B in X_1 for which B and $U_f(1)B$ are bounded in X_2, it follows that $U_f(1)$ is bounded in X_1. In this situation, Massatt [5] has recently proved the important result that point dissipative in X_2 is equivalent to compact dissipative in X_2. This is the same result as the one that is stated in Proposition 9.6 for the case when $S_f(t) = 0$ for all t. Other properties of the same type of maps have been discussed by Massatt [2].

Our primary objective is to study how the set A_f varies with f. The easiest result to obtain is the upper semicontinuity in f and is stated precisely in the following

PROPOSITION 9.7. *Suppose there is a neighborhood V of f and a bounded set $B \subset X$ such that, for each $g \in V$, $T_g(t)$ is an α-contraction, B attracts compact sets of X relative to $T_g(t)$ and $\{T_f(t)H, t \geq 0\}$ is bounded for each bounded set $H \subset X$. Then A_f is upper semicontinuous at f; that is, for any neighborhood U of A_f, there is a neighborhood W of f such that $A_g \subseteq U$ for $g \in W$.*

The idea of the proof is the following. From Proposition 9.6, A_g is compact and $A_g \subseteq \bar{B}$ for every $g \in V$. Also, Proposition 9.6 implies A_f is uniformly asymptotically stable and attracts $\bar{B}$. This is enough to complete the proof. Results similar to Proposition 9.7 for locally compact spaces have been proved by Marchetti et al. [1].

Before proceeding further, let us give some examples of semigroups which are α-contractions. Suppose A is a linear operator (bounded or unbounded) for which $-A$ is the infinitesimal generator of a strongly continuous semigroup $S(t)$ on Z. Suppose X is a Banach space which can be continuously imbedded in Z and $f\colon X \longrightarrow Z$ is a given function. The examples will be special cases of an evolutionary equation

$$\dot{u} + Au = f(u) \tag{9.4}$$

which generates a semigroup $T_f(t)$ for which the variation of constants formula holds,

$$T_f(t)x = S(t)x + \int_0^t S(t-\tau)f(u(\tau))d\tau \overset{\text{def}}{=} S(t)x + U(t)x \tag{9.5}$$

where the radius $r_e(S(t))$ of the essential spectrum of $S(t)$ satisfies

$$r_e(S(t)) \leq \exp(-\beta t), \qquad t \geq 0, \tag{9.6}$$

for some $\beta > 0$ and

(9.7) $U(t)$ is conditionally completely continuous.

From (9.6), (9.7), it follows that there is an equivalent norm with the property that $T_f(t)$ is a conditional α-contraction.

The idea for the proof is simple. Since $r_e(S(t)) \leq \exp(-\beta t)$ for any $t > 0$, the space Z can be decomposed as $Z = Z_1 \oplus Z_2$ where Z_1, Z_2 are independent of t, invariant under

$S(t)$, Z_2 is finite dimensional and the spectrum of $S(t)$ restricted to Z_2 is inside the disk centered at zero with radius $\exp(-\beta t)$ with $\beta > 0$. Now one can renorm so that $S(t)$ restricted to Z_2 has norm $\leqslant \exp(-\bar{\beta} t)$, $\bar{\beta} > 0$.

Our first example concerns the case where A is sectorial. A linear operator A on Z is *sectorial* if it is closed, densely defined, such that, for some ϕ in $(0, \pi/2)$ and $M \geqslant 1$, $a \in \mathbf{R}$, the sector

$$S_{a,\phi} = \{\lambda: \phi < |\arg(\lambda - a)| \leqslant \pi,\ \lambda \neq a\}$$

is in the resolvent set of A and $|(\lambda - A)^{-1}| \leqslant M/|\lambda - a|$ for $\lambda \in S_{a,\phi}$. A semigroup of operators $S(t)$ on Z is *analytic* if the map $t \mapsto T(t)z$ is real analytic on $0 < t < \infty$ for each $z \in Z$. If A is sectorial, then $-A$ is the generator of an analytic semigroup $S(t)$ and conversely (see Friedman [1], Henry [1], Pazy [1], Martin [1]). In addition, if A has compact resolvent, then $S(t)$ is compact for $t > 0$, $r_e(S(t)) = 0$ for $t > 0$ and so relation (9.6) in an appropriate norm is satisfied for any $\beta > 0$.

If A is sectorial, then one can define the fractional powers $(A + aI)^\alpha$ of A I aI and the spaces $Z^\alpha = D((A + aI)^\alpha)$ with the graph norm. If $f \in X^r = C^r(Z^\alpha, Z)$, $0 \leqslant \alpha < 1$, $r \geqslant 1$, then equation (9.4) generates a semigroup $T_f(t)$ on Z^α satisfying (9.5), (9.7) provided that f takes bounded sets into bounded sets and the resolvent set of A is compact. The spaces X, Y, Z are $X = Y = Z^\alpha$, $Z = Z$. To satisfy (9.6), additional restrictions must be imposed on A.

A specific example of the latter situation is

(9.8) $$\partial u/\partial t - \Delta u = f(x, u, u_x) \quad \text{in } \Omega, \qquad u = 0 \quad \text{on } \partial\Omega$$

where Ω is a bounded open set in $\mathbf{R}^n$ with smooth boundary, $f \in X^r = C^r(\Omega \times \mathbf{R} \times \mathbf{R}, \mathbf{R})$. In $W = L^2(\Omega, \mathbf{R})$, the operator $A = -\Delta$ with domain $H_0^1(\Omega) \cap H^2(\Omega) = W_0^{1,2}(\Omega) \cap W^{2,2}(\Omega)$ is sectorial with compact resolvent. For some restrictions on the rate of growth of $f(\cdot, u. v)$ in u, v, there is an α in $(0, 1)$ such that equation (9.8) generates a strongly continuous semigroup on W^α. In this case $X = W^\alpha$, $Y = \Omega \times \mathbf{R} \times \mathbf{R}$, $Z = \mathbf{R}$.

Other problems that could be considered for (9.5) are cases when $f(x, u, v)$ is independent of v; that is, $Y = \Omega \times \mathbf{R}$, or $f(x, u, v)$ independent of x, v; that is, $Y = \mathbf{R}$. As the function f is restricted to a smaller class, the generic theory will become more difficult.

One could also change the boundary conditions to some other form or consider systems of parabolic equations and obtain semigroups of the same qualitative type.

With Ω as in (9.8), $f \in X^r = C^r(\Omega \times \mathbf{R}, \mathbf{R})$, $\beta > 0$ a constant, consider the equation

(9.9) $$u_{tt} - \beta\Delta u_t - \Delta u = f(x, u) \quad \text{in } \Omega, \qquad u = 0 \quad \text{on } \partial\Omega.$$

Webb [1] has shown that (9.9) generates a semigroup $T_f(t)$ on $D(A) \times L^2(\Omega, \mathbf{R})$, $A = -\Delta$, $D(A) = H_0^1(\Omega) \cap H^2(\Omega)$. It is possible to show that $T_f(t)$ also satisfies (9.5)–(9.7) (see, for example, Massatt [4]).

In an appropriate Sobolev space, there is a semigroup generated by the beam equation

(9.10) $$\frac{\partial^2 u}{\partial t^2} + \alpha\frac{\partial^4 u}{\partial x^4} - \left(\beta + k\int_0^1 \left[\frac{\partial u(\xi, t)}{\partial \xi}\right]^2 d\xi\right)\frac{\partial^2 u}{\partial x^2} + \delta\frac{\partial u}{\partial t} = 0$$

satisfying (9.5), (9.6). Relation (9.7) does not appear to be satisfied. Ball [1], however, has obtained interesting qualitative results on this equation exploiting (9.5), (9.6) and convergence in the weak topology. This suggests that axiom (9.7) should be weakened in some way.

Retarded functional differential equations generate semigroups satisfying (9.5)–(9.7). Suppose $r > 0$, $C = C([-r, 0], \mathbf{R}^n)$, $f \in X^k = C^k(C, \mathbf{R}^n)$, $k \geq 1$ and consider the equation

$$\dot{x}(t) = f(x_t) \tag{9.11}$$

where $x_t(\theta) = x(t + \theta)$, $-r \leq \theta \leq 0$. For any $\phi \in C$, there is a unique solution $x(\theta)$ of (9.11) with initial value ϕ at $t = 0$. If we assume that $x(\phi)(t)$ is defined for $t \geq 0$ and let $T_f(t)\phi = x_t(\phi)$, $t \geq 0$, then $T_f(t)\phi$ is a strongly continuous semigroup on C. Let $S(t)$ be the semigroup on C generated by $\dot{x}(t) = 0$; that is, $S(t) = T_0(t)$,

$$S(t)\phi(\theta) = \begin{cases} \phi(t + \theta), & t + \theta < 0, \\ \phi(0), & t + \theta \geq 0. \end{cases}$$

Then $r_e(S(t)) = 0$ for $t > 0$. Also,

$$T_f(t)\phi(\theta) = S(t)\phi(\theta) + \int_0^t S(t - \tau)X_0(\theta)f(T_f(\tau)\phi)\, d\tau$$

where $X_0(\theta) = 0$ for $\theta < 0$, $= I$ for $\theta = 0$, $S(t - \tau)X_0(\theta) = 0$ for $t - \tau + \theta < 0$, $= I$ for $t - \tau + \theta \geq 0$. Thus,

$$T_f(t)\phi = S(t)\phi + \int_0^t S(t - \tau)X_0 f(T_f(\tau)\phi)\, d\tau \overset{\text{def}}{=} S(t)\phi + U(t)\phi \tag{9.12}$$

with the interpretation as above. Relation (9.12) is the analogue of (9.5) and one can show that $U(t)$ is conditionally completely continuous if f takes bounded sets into bounded sets (see Hale [3]). In this case, our spaces X, Y, Z are $X = C = Y$, $Z = \mathbf{R}^n$.

For the difference-differential equation

$$\dot{x}(t) = f(x(t), x(t - r)) \tag{9.13}$$

the spaces X, Y, Z are $X = C$, $Y = \mathbf{R}^n \times \mathbf{R}^n$, $Z = \mathbf{R}^n$. For the equation

$$\dot{x}(t) = f(x(t - r)) \tag{9.14}$$

the spaces are $X = C$, $Y = \mathbf{R}^n$, $Z = \mathbf{R}^n$.

It is possible to discuss these retarded equations in other spaces than C; for example, $\mathbf{R}^n \times L^2([-r, 0], \mathbf{R}^n)$. Relation (9.12) and $r_e(S(t)) = 0$ for $t > 0$ still remain true (see Hale [3] for references).

For the case of infinite delay in equation (9.11), it is not difficult to obtain spaces X of initial data for which one obtains a strongly continuous semigroup $T_f(t)$ on X satisfying (9.12) with $S(t)$ a strongly continuous semigroup and $U(t)$ conditionally completely continuous. For the number $r_e(S(t))$ to satisfy (9.6), one must impose some additional conditions on the space X. For example, if X is a fading memory space, the kernel approaching zero exponentially is sufficient. Many other spaces assure that (9.6) is satisfied (see Hale and Kato [1], Schumacher [1]). The survey articles of Hale [4], Corduneanu and Laksmikantham [1] should be consulted for references and more specific properties of equations with infinite delays.

Some functional differential equations of neutral type also will generate semigroups satisfying all of the properties mentioned above. For example, this will be true for the equation

$$\frac{d}{dt}\left[x(t) - \sum_{k=1}^{N} A_k x(t - r_k) - h\left(\int_{-r}^{0} A(\theta)x(t+\theta)\,d\theta\right)\right] = g(x_t) \tag{9.15}$$

with $h \in C^k(\mathbf{R}, \mathbf{R}^n)$, $g \in C^k(C, \mathbf{R}^n)$, r_k, $r > 0$, $A(\theta)$ a continuous $n \times n$ matrix, each A_k is an $n \times n$ constant matrix and the zero solution of the difference equation

$$y(t) - \sum_{k=1}^{N} A_k y(t - r_k) = 0 \tag{9.16}$$

is uniformly asymptotically stable. The semigroup $S(t)$ is the one generated by the equation

$$\frac{d}{dt}\left[y(t) - \sum_{k=1}^{N} A_k x(t - r_k)\right] = 0.$$

The parameter f in the semigroup for (9.15) is $\{A_k, A(\cdot), g, h\}$. These equations arise naturally from certain linear hyperbolic partial differential equations with nonlinear boundary conditions (see Hale [3] for references and details).

For $g = 0$ in (9.15), one also obtains semigroups generated by the functional equation

$$y(t) - \sum_{k=1}^{N} A_k x(t - r_k) - h\left(\int_{-r}^{0} A(\theta)x(t+\theta)\,d\theta\right) = 0 \tag{9.17}$$

with initial data ϕ restricted to C_0, where C_a for $a \in \mathbf{R}^n$ is defined by

$$C_a = \left\{\phi \in C: \phi(0) - \sum_k A_k\phi(-r_k) - h\left(\int_{-r}^{0} A(\theta)\phi(\theta)\,d\theta\right) = a\right\}.$$

In this case, C_0 is a closed subset of C–a nonlinear manifold which depends on the coefficients A_k and the functions A, h. From the qualitative point of view, it may be desirable not to consider (9.17), but to consider (9.14) with $g = 0$ realizing that C_a for each $a \in \mathbf{R}^n$ is invariant under the semigroup.

There is a tremendous literature on semigroups generated by evolutionary equations with delays–a combination of equations (9.4) and (9.11). Both the theory and applications are fairly well developed and it is impossible here to go into this interesting subject. A representative selection of the literature can be obtained by consulting the references in Slemrod [1], [2], Infante and Walker [1], Dafermos and Nohel [1], Fitzgibbon [1].

The above examples certainly are sufficient motivation to study in more detail semigroups $T_f(t)$ which satisfy (9.5)–(9.7). All questions, remarks, conjectures, etc. in the following pages are made for semigroups which satisfy at least these conditions. Proposition 9.2 motivates the following query.

Question 9.8. *If A_f is compact, when is there a generic set of f such that $T_f(t)$ is one-to-one on A_f?*

Note that the question is for A_f and not all of X. It does not seem to be reasonable to ask the same question for all of X.

For parabolic partial differential equations, general conditions are known which imply that $T_f(t)$ is one-to-one on all of X (see Henry [1], K. Miller [1]; for further references, see Manselli and Miller [1]). In particular, it is true if the elliptic part of the operator is the Laplacian. For retarded functional differential equations (9.11) with f analytic, $T_f(t)$ is one-to-one on A_f even though it may not be one-to-one on C. This follows because the functions defining A_f must be analytic (see Nussbaum [1]). The same result is true for neutral equations with (9.16) uniformly asymptotically stable (see Hale [3]). For other references on retarded equations, see Hale [3]. Mallet-Paret has given an example (unpublished) of a retarded functional differential equation for which $T_f(t)$ is not one-to-one on A_f.

Question 9.9. *If f is structurally stable, when is $T_f(t)$ one-to-one on A_f?*

Question 9.10. *When is A_f generically a manifold or a finite union of manifolds?*

For the case of retarded equations (9.11), defined on a compact manifold M without boundary, which are in some sense close to an ordinary differential equation, Kurzweil [1] has shown that A_f is diffeomorphic to M. Oliva [1] has generalized these results giving other conditions which imply A_f is diffeomorphic to M. Henry [1] has discussed this question for certain gradient systems of parabolic equations and shown that A_f is the union of a finite number of manifolds. We mention later special retarded functional differential equations for which A_f is the union of a finite number of manifolds.

Of course, an affirmative answer to either of the above questions requires conditions on the manner in which $T_f(t)$ depends on f and a specification of a class X^r of f's which gives enough flexibility to move the orbits in any desired direction by variations in f.

THEOREM 9.11. *If A_f is compact and, for each $t > 0$, $x \in A_f$, $D_x T_f(t)x$ is the sum of a contraction operator and a completely continuous operator, then A_f has finite Hausdorff dimension.*

For X a separable Hilbert space, this result was proved by Mallet-Paret [1]. The general case in Theorem 9.11 for X an arbitrary Banach space was proved by Mañé [2]. Ladyženskaya [1] has obtained related results for the Navier-Stokes equation. Mañé [2] proved even more—that the limit capacity of A_f is finite. The limit capacity of A_f is defined as the exponential growth rate of the number of balls of radius $\exp(-t)$ that are required to cover A_f when $t \to \infty$. The limit capacity is always at least as large as the Hausdorff dimension. Using results of Cartwright [2], [3], one obtains the following interesting consequence of Theorem 9.11.

COROLLARY 9.12. *If the conditions of Theorem* 9.11 *are satisfied, then there is an integer N such that, if there is an* $x \in X$ *such that* $T_f(t)x$ *is almost periodic, then* $T_f(t)x$ *is quasiperiodic with the number of basic frequencies being* $\leqslant N$.

Question 9.13. *For* $x \in A_f$, *when is* $T_f(t)x$ *continuously differentiable in t for* $t \in (-\infty, \infty)$?

Note again that this question pertains to A_f and not all of X.

For retarded equations (9.11), this is obviously true because $T_f(t)x$ is defined for $t \leqslant 0$. For neutral equations for which equation (9.16) is uniformly asymptotically stable and $A(\theta)$ has a continuous derivative, it is also true. The proof uses the fact that the semigroup $S(t)$ in (9.5) has essential spectrum inside the unit circle for $t > 0$ (see Hale [3]). It can be generalized to a neutral equation with the essential spectrum of $S(t)$ bounded away from the unit circle for $t > 0$ (see deOliveira [1]). Some special cases also have been considered for functional equations by Hale and deOliveira [1]. We remark that $T_f(t)x$ is differentiable in t for $x \in A_f$, but is not differentiable for every $x \in C$.

For abstract evolutionary equations, no attempt has been made to show that $T_f(t)x$ is differentiable in t for x only in A_f. One approach has been to show that $T_f(t)x$ is continuously differentiable jointly in t, x for $t > 0$ and $x \in Y$, a Banach space continuously imbedded in X. Typical hypotheses require that $T_f(t)$ is a semigroup also on a Banach space Y which can be continuously imbedded in X, Y belongs to the domain of the generator A of $T_f(t)$ and is a bounded operator from Y to X (see Dorroh and Marsden [1], Marsden and McCracken [1]). For the case where A is sectorial, Henry [1] gives more detailed results on the differentiability in t.

It should be easier to prove differentiability in t on A_f and there should be an abstract generalization of the known results for functional differential equations. It will be impossible to obtain any type of generic theory without having this type of smoothness since one cannot study the local behavior of orbits on A_f by taking a linear approximation.

Our next objective is to discuss more detailed properties of A_f and, in particular, hyperbolicity.

DEFINITION 9.14. An *equilibrium point x is hyperbolic* if there are a decomposition of X as $X = X^s \oplus X^u$ and positive constants K, α such that $X^s \cap X^u = \{0\}$, X^s, X^u are invariant under $DT_f(t)x$ and

$$|(DT_f(t)x)v| \leqslant Ke^{-\alpha t}, \quad t \geqslant 0,\ v \in X^s, \qquad |(DT_f(t)x)v| \leqslant Ke^{\alpha t}, \quad t \leqslant 0,\ v \in X^u.$$

It is implicitly assumed that $DT_f(t)x$ is defined on X^u for $t \leqslant 0$. If $T_f(t)$ is an α-contraction, then $DT_f(t)x$ is an α-contraction, the space X^u is finite dimensional and $DT_f(t)x$ is defined on X^u for $t \leqslant 0$.

Suppose $\Gamma = \{T_f(t)x,\ 0 \leqslant t < \omega\}$ is an ω-periodic orbit for which $T_f(t)x$ is continuously differentiable in t. (This is one reason we need an answer to Question 9.13.) Let H be a hyperplane in X; that is, $H = \{y \in X \colon h(y) = 0$ where h is a nontrivial continuous linear functional on $X\}$. Suppose there are a neighborhood W of zero in H and a function $\phi \colon W \longrightarrow X$ such that $\phi(0) = x$ and $\phi(W)$ is a C^1-submanifold of codimension one. Let M

be the tangent space to $\phi(W)$ at x. The set $V = \phi(W)$ is said to be a *transversal* to Γ at x if $dT_f(t)x/dt$ does not belong to M. Let π be the Poincaré map defined on a neighborhood U of x in a transversal V to Γ at x. Then πy is continuously differentiable for $y \in U$.

DEFINITION 9.15. Suppose the notation is as in the previous paragraph. A *periodic orbit* Γ *is hyperbolic* if there are a decomposition of M as $M = M^s \oplus M^u$ and positive constants K, α such that $M^s \cap M^u = \{0\}$, M^s, M^u are invariant under $D\pi(x)$ and

$$|(D\pi(x))^n v| \leqslant Ke^{-\alpha n}, \quad n \geqslant 0,\ v \in M^s, \quad |(D\pi(x))^n v| \leqslant Ke^{\alpha n}, \quad n \leqslant 0,\ v \in M^u.$$

One can define the stable and unstable manifolds for hyperbolic equilibrium points and periodic orbits in the usual way. For functional differential equations (see Hale [3]) and nonlinear evolutionary equations (9.4) with A sectorial (see Henry [1]), the unstable and stable manifold intersect transversally at the point or orbit, the unstable manifold is an immersed submanifold of finite dimension and the local stable manifold is diffeomorphic locally to the stable manifold of the linear approximation. In general, it is not possible to define a global stable manifold because the solution operator $T_f(t)$ is not one-to-one for negative t when it is defined. An abstract generalization of these results is needed.

DEFINITION 9.16. Suppose $T_f(t)x$ is continuously differentiable in t if x belongs to a periodic orbit. Then $T_f(t)$ is said to be *Kupka-Smale* if:

(i) All equilibrium points are hyperbolic.

(ii) All periodic orbits are hyperbolic.

(iii) The local stable and global unstable manifolds of equilibrium points and periodic orbits intersect transversally.

Question 9.17. *Is the set of* $f \in \mathcal{X}^r$ *corresponding to Kupka-Smale semigroups generic in* $\mathcal{X}^r$?

Some results have recently been obtained. For the general retarded functional differential equations (9.11) with $f \in \mathcal{X}^r$, the set of C^r-functions from $C([-h, 0], \mathbf{R}^n)$ to $\mathbf{R}^n$, Mallet-Paret [2], [3] has shown the answer to the question is affirmative. If $f \in C^r(\mathbf{R}^n \times \mathbf{R}^n, \mathbf{R}^n)$ as in (9.13), he has also shown the same result is true. If $f \in C^r(\mathbf{R}^n, \mathbf{R}^n)$ as in (9.14), the answer to the question is not known. The fundamental part of the proofs always involves (as was the case in finite dimension) showing that the set of f for which the periodic orbits with one not in the spectrum of $D\pi(x)$ is dense.

For the neutral equations discussed above, deOliveira [1] has shown that the f such that the equilibrium points are hyperbolic is open and dense. The general KS property has not been discussed.

For the parabolic equation

$$u_t = u_{xx} + f(x, u), \qquad 0 < x < 1, \tag{9.18}$$

$$\alpha u(t, 0) + \beta u_x(t, 0) = 0, \qquad \gamma u(t, 1) + \delta u_x(t, 1) = 0, \tag{9.19}$$

the ω-limit set of every bounded orbit is a single equilibrium point; that is, a solution of

$$u_{xx} + f(x, u) = 0 \tag{9.20}$$

satisfying the boundary conditions at $x = 0, x = 1$ (see Matano [2]). It is not difficult to show that these equilibrium points are hyperbolic generically in $f \in C^1(\mathbf{R} \times \mathbf{R}, \mathbf{R})$. The same result has recently been proved by Brunovsky and Chow [1] for the equation

$$u_t = u_{xx} + f(u), \qquad 0 < x < 1, \tag{9.21}$$

with Neumann or Dirichlet boundary conditions, where the function $f \in C^2(\mathbf{R}, \mathbf{R})$ with the Whitney topology is independent of the space variable x. The restriction to this smaller class of functions makes the result nontrivial. The methods appear to apply to the general boundary conditions (9.19).

To see why this is nontrivial, consider the Neumann boundary condition $u_x(0) = u_x(1) = 0$ for the equilibrium equation (9.20). If $\omega(b, f)$ is the period of the solution of (9.20) through the point $(u(0), u_x(0)) = (b, 0)$, then one must show that the solutions b of the equation $\omega(b, f) - 2 = 0$ satisfy $\partial\omega(b, f)/\partial b \neq 0$ for a residual set of $f \in C^2(\mathbf{R}, \mathbf{R})$. There are classical examples of nonlinear functions f where $\omega(b, f)$ is a constant function of b (see, for example, Urabe [1]). This implies that the derivatives of $\omega(b, f)$ with respect to b are very complicated functions of the derivatives of f.

For the equation

$$u_t = \Delta u + f(u) \quad \text{in } \Omega, \qquad \alpha u + \beta \partial u/\partial n = 0 \quad \text{on } \partial\Omega \tag{9.22}$$

where Ω is a bounded open connected set in $\mathbf{R}^n$ with smooth boundary, the corresponding result is not known. In this case, it is reasonable to consider the parameters as f and Ω and to prove that the equilibrium points are hyperbolic generically in (f, Ω). An important role should be played by results similar to the ones of Uhlenbeck [1] on the simplicity of the eigenvalues of the Laplacian operator generically in Ω.

To the author's knowledge, property (iii) in the definition of Kupka-Smale systems has not been discussed for the general equation (9.22). Some partial results of Henry [1], [2] are mentioned below.

DEFINITION 9.18. Let $\Omega(f)$ be the nonwandering set of $T_f(t)$. A semigroup $T_f(t)$ is said to be *Morse-Smale* if $\Omega(f)$ is the union of a finite number of equilibrium points and periodic orbits, each hyperbolic with stable and unstable manifolds intersecting transversally.

Question 9.19. *Are Morse-Smale systems open and structurally stable*?

One of the first important problems in any attempt to discuss these last general questions is to construct several examples. The first step would be to consider systems which have properties analogous to gradient systems for ordinary differential equations.

Gradient flows defined by parabolic equations have been analyzed by Henry [1], [2]. In particular, it is shown that the equation

$$\begin{aligned} &u_t = u_{xx} + \lambda(u - u^3), \qquad 0 < x < \pi, \ t \geq 0, \\ &u = 0 \quad \text{at } x = 0, \pi \end{aligned} \tag{9.23}$$

is structurally stable in $H_0^1(0, 1)$ for $\lambda \in (0, 9)$, $\lambda \neq 1$, $\lambda \neq 4$, A_λ is a point for $0 < \lambda < 1$,

A_λ is a smooth closed arc with three equilibrium points for $1 < \lambda < 4$, A_λ is 2-dimensional and contains 5 equilibrium points for $4 < \lambda < 9$.

It is not difficult to see why this should be true if we use our detailed knowledge of the maximal compact invariant set. In fact, for any $\lambda > 0$, one can use the invariance principle and the function

$$V = \int_0^\pi \frac{1}{2}u_x^2 - \lambda\left(\frac{u^2}{2} - \frac{u^4}{4}\right)$$

to show that every solution is bounded and approaches a solution of

$$u_{xx} + \lambda(u - u^3) = 0 \tag{9.24}$$

with $u(0) = u(\pi) = 0$. For each $\lambda = 0$, this equation with these boundary conditions has a finite number of solutions $\{u_j, j = 1, 2, \ldots, n(\lambda)\}$ with unstable manifolds $W^u(u_j)$ of finite dimension. One can show that this implies A_λ is compact and attracts bounded sets, $A_\lambda = \bigcup_{j=1}^{n(\lambda)} W^u(u_j)$, and is upper semicontinuous in λ. If $\lambda \in (0, 1)$, $n(\lambda) = 1$, $u_1 = 0$, $A_f = \{0\}$. For $\lambda = 4$, the half-period for the solutions of the linear part of (9.24) is π. For $\lambda = 4$ and close to 4, there are two additional solutions u_2, u_3 near zero, the solution $u = 0$ is unstable with dim $W^u(0) = 1$. Thus, $n(\lambda) = 3$ and $A_\lambda = W^u(0) \cup \{u_2\} \cup \{u_3\}$ is a smooth closed arc. At $\lambda = 9$, there is another bifurcation at zero in a direction independent of the previous bifurcation. This yields two more solutions–each unstable–and the two dimensional set A_λ mentioned before.

For retarded functional differential equations, it does not seem possible to have a system which is a gradient system. However, there are some equations which have some of the same qualitative properties. We discuss such an example in some detail since it brings out many of the difficulties encountered in infinite dimensions and it also leads to many interesting and specific unsolved problems.

Consider the equation

$$\dot{x}(t) = -\int_{-1}^0 a(-\theta)g(x(t + \theta))\,d\theta \tag{9.25}$$

where $g \in C^2(\mathbf{R}, \mathbf{R})$, $a \in C^2([0, 1], \mathbf{R})$, $\int_0^x g \to \infty$ as $|x| \to \infty$, $a(1) = 0$, $a(s) \geqslant 0$, $\dot{a}(s) \leqslant 0$, $\ddot{a}(s) \geqslant 0$.

Under these hypotheses, every solution of (9.25) is defined and bounded for $t \geqslant 0$. If $C = C([-1, 0], \mathbf{R})$, then the semigroup $T_{a,g}(t)\colon C \to C$, $t \geqslant 0$, is well defined by the relation $T_{a,g}(t)\phi(\theta) = x(\phi)(t + \theta)$, $-1 \leqslant \theta \leqslant 0$, where $x(\phi)$ is the solution of (9.25) through ϕ. If there is an $s \in [0, 1]$ such that $\ddot{a}(s) > 0$, then the ω-limit set of any solution is an equilibrium point of (9.25); that is, a zero of g. If $\ddot{a}(s) = 0$ for all $s \in [0, 1]$ (that is, a is linear), then the ω-limit set of any solution is either an equilibrium point or a closed orbit given by $\{p_t, 0 \leqslant t \leqslant 1\}$, where $p(t)$ is a one-periodic solution of the ordinary differential equation

$$\ddot{y} + a(0)g(y) = 0. \tag{9.26}$$

For a proof of these results, see Hale [3, p. 122].

Let us consider first the case where $\ddot{a}(s) > 0$ for all $s \in [0, 1]$ and g has a finite number of simple zeros $\{\alpha_j, j = 1, 2, \ldots, k\}$. Let

$$W^s(\alpha_j) = \{\phi \in C\colon T_{a,g}(t)\phi \longrightarrow \text{the constant function } \alpha_j \text{ as } t \longrightarrow \infty\},$$

$$W^u(\alpha_j) = \{\phi \in C\colon T_{a,g}(t)\phi \longrightarrow \text{the constant function } \alpha_j \text{ as } t \longrightarrow -\infty\}.$$

It is not difficult to show that each α_j is hyperbolic and $\dim W^u(\alpha_j) \leqslant 1$. Then

$$A_{a,g} = \bigcup_{j=1}^{k} W^u(\alpha_j)$$

is a union of a finite number of manifolds and $T_{a,g}(t)$ is one-to-one on $A_{a,g}$.

It is possible to construct open sets $\mathcal{A} \subset C^2([0, 1], \mathbf{R})$, $\mathcal{G}$ in the Whitney topology on $C^2(\mathbf{R}, \mathbf{R})$ such that $A_{a,g}$ is upper semicontinuous at each $(a, g) \in \mathcal{A} \times \mathcal{G}$. The way to do this can be discovered from the proof of the above results in Hale [3] and the method of proof of Proposition 9.7. We can now prove

PROPOSITION 9.20. *If $\ddot{a}(s) > 0$ for all $s \in [0, 1]$, the zeros $\{\alpha_j, j = 1, 2, \ldots, k\}$ of g are simple, $(a, g) \in \mathcal{A} \times \mathcal{G}$, and the ω-limit set of each $W^u(\alpha_j) \setminus \{\alpha_j\}$ is a stable equilibrium, then (a, g) is structurally stable.*

PROOF. Since $A_{a,g}$ is upper semicontinuous and uniformly asymptotically stable, one can choose a neighborhood U of $A_{a,g}$ (as small as desired) and a neighborhood V of (a, g) in $\mathcal{A} \times \mathcal{G}$ such that $T_{b,h}(t)U \subset U$, $t \geqslant 0$, $(b, h) \in V$. Since the zeros of g are simple, we have $\dim W^u(\alpha_j) \leqslant 1$.

The assumption that the ω-limit set of each $W^u(\alpha_j)\setminus\{\alpha_j\}$ is a stable equilibrium implies $A_{a,g}$ is one dimensional as shown in Figure 17. The theory of the neighborhood of a saddle point in Hale [3] implies that the flow in U is given as shown in Figure 18, where the vertical manifolds have codimension 1 and $A_{a,g}$ has dimension one. This implies that $A_{b,h}$ has the same topological structure as $A_{a,g}$ as well as the same type of flow.

FIGURE 17.

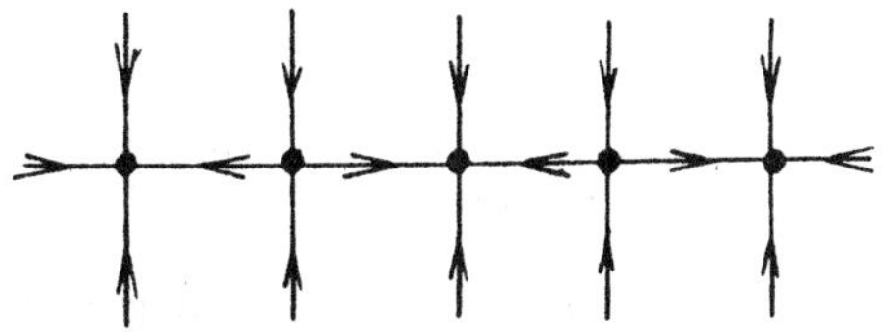

FIGURE 18.

One can now ask the following interesting question.

Question 9.21. *Suppose a is a fixed function satisfying $\ddot{a}(s) > 0$ for $s \in (0, 1)$. Let $\mathcal{G}_k = \{g$ which have exactly $2k + 1$ zeros which are all simple$\}$. How many different connected components of structurally stable systems are in $\mathcal{G}_k$?*

If $k = 0$, that is, g has only one zero α, then $A_{a,g} = \{\alpha\}$, the constant function α and all $g \in \mathcal{G}_0$ are structurally stable and there is only one component.

If $k = 1$, then $g \in G_1$ has three simple zeros, $\alpha_1 < \alpha_2 < \alpha_3$, with α_1, α_3 asymptotically stable and α_2 a saddle point. Since the unstable manifold at α_2 is smooth, one dimensional and $A_{a,g}$ is uniformly asymptotically stable, it follows that $A_{a,g}$ is a one dimensional manifold with boundary points α_1, α_3. Again, all elements of G_1 are structurally stable.

The topological structure of $A_{a,g}$ is not understood for the case when g has given zeros $\alpha_1 < \alpha_2 < \alpha_3 < \alpha_4 < \alpha_5$. If we consider g depending on some parameters μ with $g(x, \mu) = x^5$ at $\mu = 0$, then $A_{a,g}(\mu)$ is $\{0\}$ for $\mu = 0$. There is a center manifold in a neighborhood of $x = 0$ which is one dimensional and smooth in μ. As μ varies near $\mu = 0$, $g(x, \mu)$ can have as many as 5 zeros which must lie on this center manifold. Thus, $A_{a,g}(\mu)$ is either a point or a one dimensional manifold with boundary for μ small. As μ increases, it is conceivable that the topological structure of $A_{a,g}$ changes. Let us give some intuitive reasons for why this is possible. The author wishes to acknowledge conversations with John Mallet-Paret and Shui-Nee Chow which were of great assistance in the remaining discussion of this section.

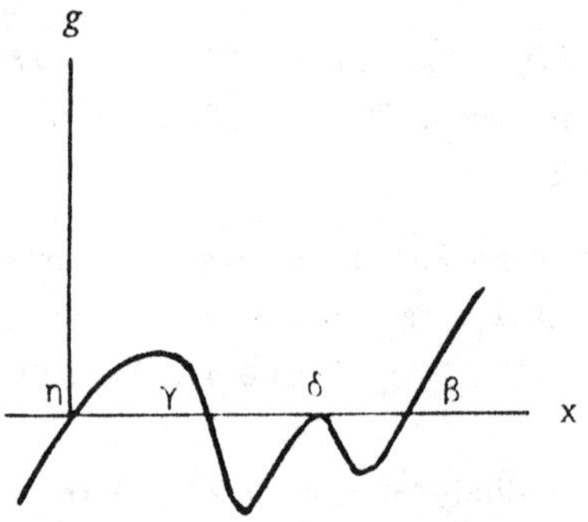

FIGURE 19.

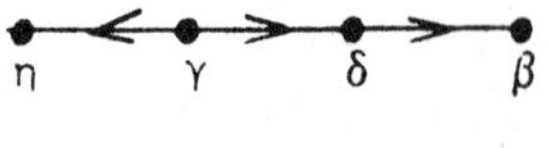

FIGURE 20.

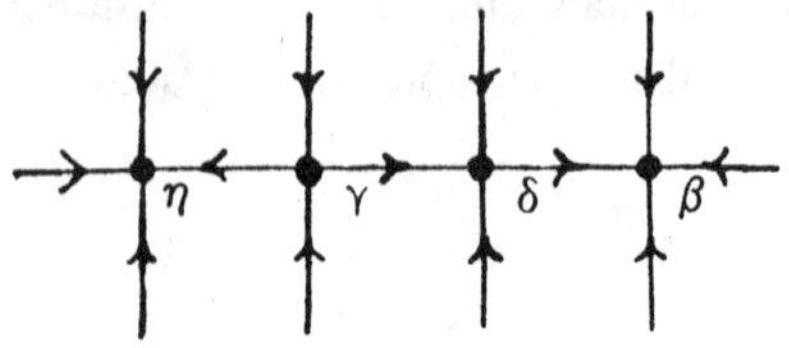

FIGURE 21.

Suppsoe g has five zeros and the general shape shown in Figure 19. If a is strictly convex and very close to the δ-function at zero, then $A_{a,g}$ is shown in Figure 20. The complete flow near $A_{a,g}$ is shown in Figure 21 with the exponential decay toward $A_{a,g}$ being very rapid and much greater than the convergence of the flow on $A_{a,g}$ toward β.

If $g'(\beta) = \alpha$, then the flow near β is determined by the roots of the characteristic equation

$$\lambda = -\alpha \int_{-1}^{0} a(-\theta) e^{\lambda\theta} \, d\theta. \tag{9.27}$$

One can show there are an α_0 and a strictly convex function a_0 with the property that the roots of this equation with maximum real part corresponds to a double root. Change g slightly so that the zero δ disappears and g has only three zeros. Because of the nature of a stable node with both eigenvalues equal, there are two distinguished directions along which β can be approached along the unstable manifold of γ as shown in Figure 22. Choose one

FIGURE 22.

of these directions. Now, one should be able to move g back so that the double zero δ appears but the unstable manifold of δ approaches β from the other distinguished direction. If this is the case, then the new $A_{a,g}$ is shown in Figure 23. Now one can further change g

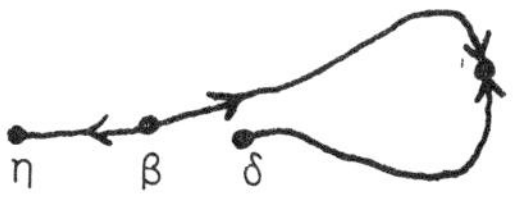

FIGURE 23.

to make δ become two distinct zeros δ_1, δ_2 and the resulting $A_{a,g}$ is structurally stable. Thus, there are at least two distinct connected components if g has five zeros. To go from one of these classes to the other keeping the zeros of g simple, it is not difficult to see that one must have a saddle connection at some point. Since we have not verified that it is possible to move g as specified above, we state this as

CONJECTURE 9.22. *There is a saddle connection for some (a, g) with a strictly convex and g having five simple zeros. The set of (a, g) for which this is true is not generic.*

Another interesting conjecture about equation (9.24) is the following

CONJECTURE 9.23. *If $\ddot{a}(s) > 0$ for $s \in [0, 1]$ and the zeros of g are simple, then the set $W = \{W^s(\alpha_j)$: α_j is asymptotically stable$\}$ is dense in C.*

One could try to prove this using the same idea that Henry **[1]** employed to prove the same result for equation (9.23). Let α be a saddle point and for any $r > 0$, $\tau > 0$, let $W^s_{\tau,r}(\alpha) = \{\phi \in C\colon T_{a,g}(\tau)\phi \in W^s(\alpha), |T_{a,g}(\tau)\phi - \alpha| \leq r\}$. The set $W^s_{\tau,r}(\alpha)$ is closed. If it contains a ball $B = \{\phi_1 + ph\colon |h| = 1, 0 \leq p \leq r_0\}$, then $T_{a,g}(\tau)B$ has been flattened so that it is contained in a submanifold of codimension one. In particular, $\partial T_{a,g}(t)(\phi_1 + ph)/\partial p$ at $p = 0$ is in the tangent space of $W^s(\alpha)$. This function $D_{\phi_1}T_{a,g}(t)h$ is the solution of the linear variational equation

$$\dot{y}(t) = -\int_{-1}^{0} a(-\theta)g'(T_{a,g}(t)\phi_1)y(t)$$

with $y_0 = h$. If the solution operator of the adjoint of this equation is one-to-one, then the assertion that $W^s_{\tau,r}(\alpha)$ contains a ball is false. In fact, take a vector $\eta \neq 0$ such that $\langle \eta, \phi \rangle = 0$ for all ϕ in the tangent space to $W^s(\alpha)$ at $T_{a,g}(\tau)\phi_1$. Integrating the adjoint equation for a function z_t on $[0, \tau]$ with $z_\tau = \eta$, one obtains, for all h, $(z_0, h) = (z_\tau, h) = (\eta, D_{\phi_1} T_{a,g}(\tau)h) = 0$. Thus, $z_0 = 0$. Since the solution operator for the adjoint equation is one-to-one, it follows that $\eta = 0$. Consequently, the set $W^s_{\tau,r}(\alpha)$ contains no ball, which implies that $W^s_{\tau,r}$ is closed and nowhere dense. Thus, $W^s(\alpha)$ has no interior. This will prove the conjecture.

10. Bifurcation in infinite dimensions

For systems of parabolic partial differential equations, retarded functional differential equations and certain types of neutral equations, several problems of local bifurcation near an equilibrium point can be discussed in essentially the same way as for ordinary differential equations. In particular, the results in §4 on bifurcation at a zero eigenvalue and in §5 on bifurcation at a focus extend verbatum (see, for example, Marsden and McCracken [1], deOliveira and Hale [1], Kielhöfer [1]). The center manifold theorem plays an important role in the theoretical justification of the results. For illustrations of the results in §8 for the case of a double eigenvalue zero, see Carr [1], Howard [1].

Very little attention has been devoted to the manner in which local bifurcation influences the global flow defined by the equations. In particular, how does the maximal compact invariant set change at bifurcation points?

Marchetti et al. [1] have some results for locally compact spaces. More specifically, they relate the change in stability of A_f to the appearance of new invariant sets. It would be interesting to extend this to Banach spaces for the type of semigroups mentioned in the previous section. For the gradient equation (9.23), we have discussed in the previous section the results of Henry [1], [2] on the manner in which the maximal compact invariant set changes with a parameter.

For the more general equation (9.21) with f depending only on u and boundary conditions (9.19), Brunovsky and Chow [1] have shown the following interesting fact: generically in $f \in C^2(\mathbf{R}, \mathbf{R})$, the bifurcations always are saddle-node type; that is, a saddle and node coalesce and disappear. In particular, this has the following interesting implication for second order ordinary differential equations. The period $\omega(b, f)$ of the solution of $u_{xx} + f(u) = 0$ through the point $u(0) = a$, $u_x(0) = 0$ is a Morse function generically in f. Smoller and Wasserman [1] have also discussed this function in detail for specific functions f.

For the equation $\dot{x}(t) = -f(x(t-1))$, where f is piecewise linear with a finite number of jump discontinuities, Walther [1] and Chow and Walther [1] have discussed the set A_f and the dynamics on A_f in some detail. The hypotheses on f serve as a model for the equation of Wright (see Nussbaum [2]).

For equation (9.25), we discussed how the set $A_{a,g}$ could bifurcate by the creation of an arc connecting two saddle points. This is always a bifurcation caused by the global properties of the flow.

The analogue of the Generic Hopf Bifurcation Theorem (Theorem 5.6) for infinite dimensional systems is prevalent in the literature. It seems reasonable to say that it is the generic situation. On the other hand, we must be careful because there are so many different ways to model a problem in infinite dimensions. For example, in equation (9.25), the vector field should be considered determined by the function $a(s)g(x)$, $s \in [0, 1]$, $x \in \mathbf{R}$; that is, as a product of a function of s and a function of x. This is certainly an easier model to discuss than one which uses a general function $h(s, x)$. However, it is still feasible that all interesting qualitative behavior of solutions can be obtained by considering the integrand as a product of two functions. For equation (9.25), the function a_0, g, satisfying $xg(x) > 0$ for $x \neq 0$, $g'(0) = 1$, $a_0(s) = 4\pi^2(1 - s)$, is a bifurcation point for every g. The linear variational equation at zero has all eigenvalues with negative real parts except the purely imaginary ones $\pm 2\pi i$. Following the procedure in §5, one can compute the bifurcation function $G(b, a, g)$ for the periodic orbits near zero with the amplitude of the periodic orbits being given approximately by b. For the Generic Hopf Bifurcation Theorem to be applicable, one should have $G(b, a_0, g) = \beta_{a_0g}b^3 + o(|b|^3)$ as $|b| \longrightarrow 0$, $\beta_{a_0g} \neq 0$. However, one can show that $\beta_{a_0,g} = 0$ for all g in the above class! Thus, the bifurcation from the focus is never a bifurcation of order one. This could not happen if $a(s)g(x)$ were replaced by a general function $h(s, x)$. For a detailed discussion of this point as well as the nature of the bifurcation point, see Hale [7].

In the previous section, we noted that it was not known if the Kupka-Smale systems were generic for $\dot{x}(t) = f(x(t - 1))$. This is a problem of the same type as in the previous paragraph where severe restrictions are imposed on the vector field. We have seen before that the same difficulties occur in parabolic equations when the nonlinearities depend only on the dependent variable.

Bifurcation of a periodic orbit to a torus for infinite dimensional systems has been discussed in some detail (see Iooss and Joseph [1] for references). Holmes and Marsden [1] show that equation (9.10) with a periodic forcing can have homoclinic points.

There are some bifurcation problems that are unique to infinite dimensions. For example, consider the scalar parabolic equation

$$(10.1) \qquad u_t = \Delta u + f(u) \quad \text{in } \Omega, \quad \partial u/\partial n = 0 \quad \text{on } \partial\Omega$$

where Ω is a bounded open connected set in $\mathbf{R}^n$ with smooth boundary $\partial\Omega$. If $W = L^2(\Omega, \mathbf{R})$, suppose this equation generates a strongly continuous semigroup on W^α, $\alpha \in (0, 1)$. Then every bounded orbit is precompact and the ω-limit set of any bounded orbit can be shown to belong to the set E of equilibrium solutions of (10.1); that is, the set of solutions of the equation

$$(10.2) \qquad \Delta u + f(u) = 0 \quad \text{in } \Omega, \quad \partial u/\partial n = 0 \quad \text{on } \partial\Omega$$

(see Matano [1]). For $n = 1$, Matano [2] has shown that the ω-limit of any bounded orbit is a single point in E. The same result for arbitrary n is contained in Hale and Massatt [1] with the additional restriction that every point in a connected component of E (which is not a single point) has zero as a simple eigenvalue of the corresponding linear variational equation. In particular, if all solutions are bounded and the set of equilibrium points is finite, then A_f is compact, $A_f = \bigcup_j W^u(\alpha_j)$ where $W^u(\alpha_j)$ is the unstable manifold of the equilibrium point α_j.

For Ω convex, Casten and Holland [1], Matano [1] have shown that the only stable equilibrium points of (10.1) are spatially homogeneous; that is, they correspond to constant functions–the zeros of f. Chafee [2] previously has proved this result for $n = 1$. Bardos, Matano and Smoller [1] have proved a similar result for one equation of the form (10.1) coupled with some ordinary differential equations. Casten and Holland [1], Matano [1] also have proved this type of instability for some other special domains Ω.

If $n \geqslant 2$, $f(u) = u - u^3$ (or any function with similar properties), Matano [1] has shown that there are stable equilibrium points of (10.1) which are not spatially homogeneous if the region Ω has certain properties. For example, it is sufficient to have Ω consist of disjoint open convex connected sets Ω_1, Ω_2 joined by a channel Ω_3 which is not too wide compared to the size of Ω_1, Ω_2 (see Figure 24).

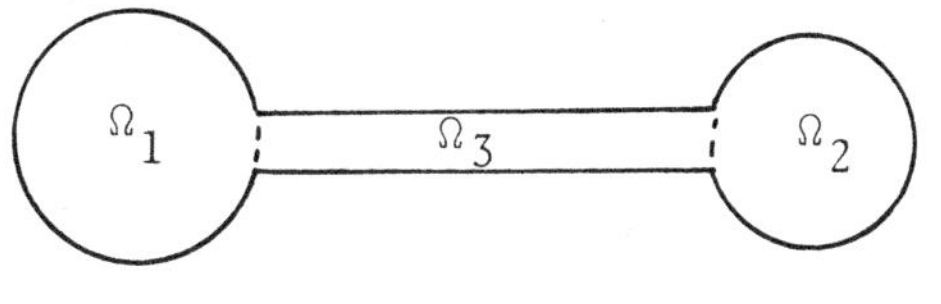

FIGURE 24.

This leads to the following interesting bifurcation problem. Let $f(u) = \lambda u - u^3$, $\lambda \in \mathbf{R}$, and let Ω_ϵ, $\epsilon \in \mathbf{R}$, be a simply connected open set in $\mathbf{R}^n$ with smooth boundary, Ω_0 convex. Let $T_{\lambda,\epsilon}(t)$ be the semigroup generated by (10.1) for this f and Ω_ϵ. Then $T_{0,0}(t)$ has a unique equilibrium point $u = 0$ which is uniformly asymptotically stable and attracts bounded sets of W^α. The maximal compact, invariant set $A_{\lambda,\epsilon}$ for $(\lambda, \epsilon) = (0, 0)$ is shown in Figure 25a. For $\lambda > 0$ and small, there are only three equilibrium points $0, \pm\lambda^{1/2}$ and $A_{\lambda,0}$ as shown in Figure 25b. Now suppose that the region Ω_ϵ with increasing ϵ becomes nonconvex so that the second eigenvalue $\mu_2(\epsilon)$ for the Neumann problem for the Laplacian satisfies $f'(0) = \lambda < \mu_2(\epsilon)$ for $0 \leqslant \epsilon < \epsilon_0$, $f'(0) = \mu_2(\epsilon_0)$ and $f'(0) > \mu_2(\epsilon)$ for $\epsilon > \epsilon_0$. The origin becomes a bifurcation point at $\epsilon = \epsilon_0$, and $A_{\lambda,\epsilon}$ for $\epsilon > \epsilon_0$ in a neighborhood of ϵ_0 is shown in Figure 25c. The solutions bifurcating from 0 are spatially nonhomogeneous. Suppose that Ω_ϵ as $\epsilon \longrightarrow \infty$ approaches the set $\Omega_\infty = \Omega_1 \cup \Omega_2 \cup L$ where Ω_1, Ω_2, L are disjoint, Ω_1, Ω_2 are open convex connected sets and L is an $(n - 1)$ dimensional closed manifold of codimension 1 joining Ω_1 to Ω_2 (in $\mathbf{R}^2$, an arc). There will be some point ϵ_1 such that the conditions for Matano's theorem are satisfied for $\epsilon > \epsilon_1$. There will be a stable spatially nonhomogeneous equilibrium point. Thus, the set $A_{\lambda,\epsilon}$ has had to undergo another bifurcation. This could not have occurred at the equilibrium points $0, \pm\lambda^{1/2}$ since $\pm\lambda^{1/2}$ are uniformly asymptotically stable and 0 has a two dimensional unstable manifold. Thus, it is reasonable to conjecture that the unstable spatially nonhomogeneous solutions undergo a bifurcation as shown in Figure 25d at least if the family of regions Ω_ϵ are chosen appropriately. The bifurcation diagram is conjectured to be the one shown in Figure 25e for certain Ω_ϵ. All the solutions are spatially nonhomogeneous except $0, \pm\lambda^{1/2}$ and the stable ones occur as a secondary bifurcation. It would be interesting to obtain these curves numerically. Hale and Vegas [1] showed there is an $\epsilon_0 > 0$ such that the bifurcation diagram for $\epsilon > \epsilon_0$ is the one shown in Figure 25e provided that the region Ω_ϵ satisfies some conditions in ϵ. The most interesting condition is that the third eigenvalue of the Laplacian on Ω_ϵ is bounded away from zero for all ϵ. This restriction arises because the analysis consists in treating the problem as a bifurcation from a double eigenvalue at $\epsilon = 0$.

FIGURE 25a.

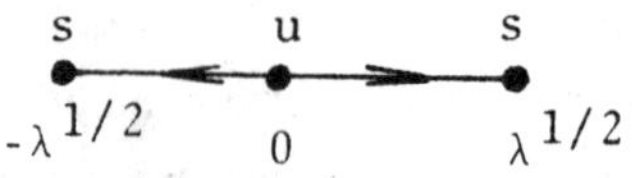

FIGURE 25b.

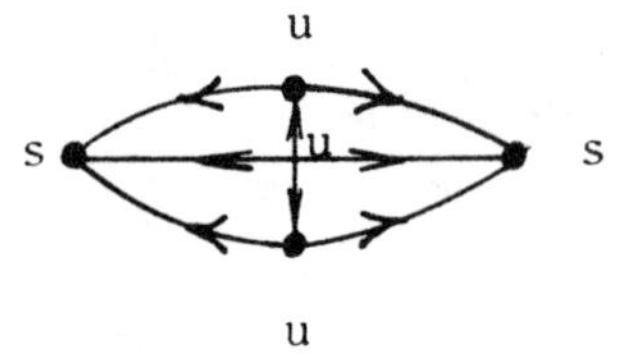

FIGURE 25c.

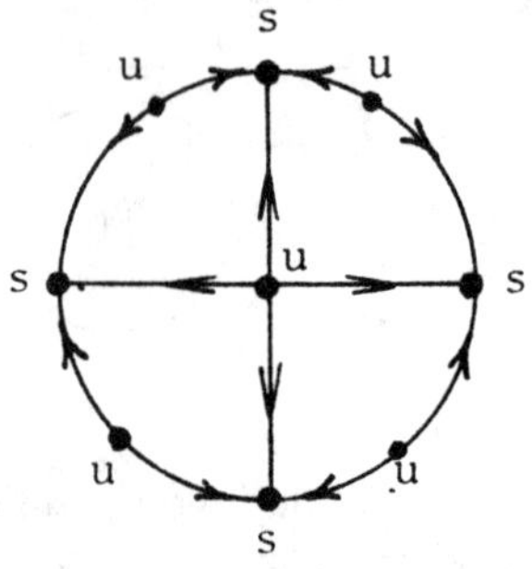

FIGURE 25d.

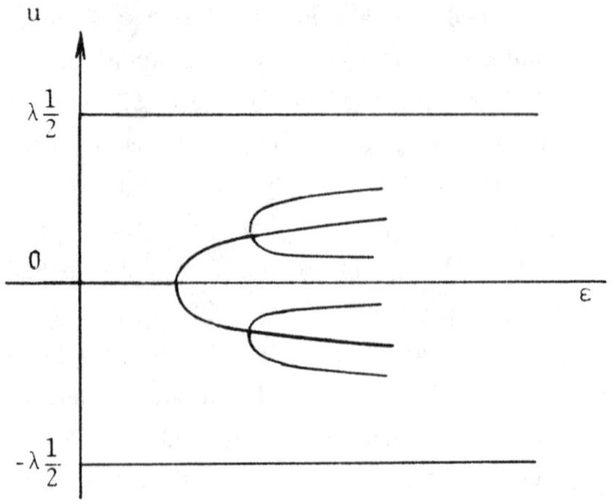

FIGURE 25e.

If the bifurcation diagram is the one depicted in Figure 25e and one introduces another parameter in the vector field f, then one should be able to break all of the symmetries in the problem and obtain a jumping from one state to another as in buckling problems.

Chafee and Infante [1] have investigated the Dirichlet problem

$$u_t = \Delta u + \lambda f(u) \quad \text{in } \Omega, \qquad u = 0 \quad \text{on } \partial\Omega \tag{10.3}$$

for $\Omega = [0, \pi]$, $\lambda \in \mathbf{R}$. Under suitable conditions on f including $uf(u) > 0$ for $u \neq 0$, they show that the only stable equilibrium solution is $u = 0$ if $0 < \lambda < \lambda_1$, and either of two functions $u_\pm(\lambda)$ for $\lambda > \lambda_1$ where $u_+(\lambda)$ is positive on $(0, \pi)$ and $u_-(\lambda)$ is negative on $(0, \pi)$. For $\Omega \subseteq \mathbf{R}^n$, one should encounter a much more interesting type of behavior by changing the shape of the region Ω as we did with the Neumann boundary conditions. To the author's knowledge, this problem has not been investigated.

It would also be interesting to discuss the same equation with mixed boundary conditions varying both the boundary conditions and the region.

For systems of reaction diffusion equations with one space variable and Neumann boundary conditions, one can also obtain stable equilibrium solutions which are not spatially homogeneous. This is a bifurcation problem where one creates an instability of the zero solution by making the linear approximation have a zero eigenvalue by varying the diffusion coefficients and the linear coupling terms (for references and a detailed discussion, see Fife [1]). This corresponds to primary bifurcation from the trivial solution whereas the stable solution in the previous discussion was created through a secondary bifurcation.

For reaction diffusion equations in an unbounded domain, there are some very interesting bifurcation problems associated with traveling wave solutions. Due to limitations in space, we can only refer to Fife [1] for references.

Another problem which becomes important in infinite dimensions is the discussion of semigroups $T_\lambda(t)$ which are continuous in a parameter λ, but are not continuously differentiable in λ. In particular, it is possible to discuss how invariant sets change with λ, how the stable manifolds change with λ, etc.? In the qualitative theory, these are fundamental questions. If $T_\lambda(t)$ has a maximal compact invariant set A_λ, then all of the interesting properties of the orbits of $T_\lambda(t)$ are determined by the behavior of the orbits on A_λ. The orbits on A_λ are bounded and defined on $(-\infty, \infty)$. As we indicated earlier, for some important types of semigroups, the fact that orbits on A_λ are defined on $(-\infty, \infty)$ implies that $T_\lambda(t)x$ for $x \in A$ is continuously differentiable in t; that is, the elements in A_λ are "smoother" than a generic element of the underlying space X. Can this additional smoothness on A_λ be exploited to discuss how A_λ varies with λ? For differential difference equations with the parameter λ being the delays, we indicate why this is feasible. No corresponding results have been obtained for partial differential equations.

The following lemma is crucial to the analysis. It asserts that the fixed points of a map can be shown to be continuously differentiable in a parameter λ by requiring only the derivative of the map be differentiable with respect to λ on the fixed point set. This lemma was stated in Hale [3] with the omission of the obviously necessary fact that the derivative in (iv) below is continuous.

LEMMA 10.1. *Let F be a closed subset of a Banach space X,* Int $F \neq \emptyset$, *Λ be an open subset of a Banach space Y, where* Int F *denotes the interior of F. Assume that T: $F \times \Lambda \longrightarrow F$, $(x, \lambda) \longrightarrow T(x, \lambda)$, satisfies the following set of hypotheses:*

(i) *$T(x, \cdot)$: $\Lambda \longrightarrow F$ is continuous;*

(ii) *$T(\cdot, \lambda)$: $F \longrightarrow F$ is continuous and has, for each λ, a unique fixed point $x(\lambda)$ which is continuous in λ;*

(iii) *if $x(\Lambda) = F_1$, then $T(x, \lambda)$ is continuously differentiable in λ for $(x, \lambda) \in F_1 \times \Lambda$;*

(iv) *there are an open set $F_1 \subset X$, $F \subset F_2$, and a $\delta \in [0, 1)$ such that the derivative of $T(x, \lambda)$ with respect to x is continuous and has norm $< \delta$ for all $(x, \lambda) \in F_2 \times \Lambda$.*

Then the fixed point $x(\lambda)$, $\lambda \in \Lambda$, of $T(x, \lambda)$ is continuously differentiable in λ.

PROOF. The proof is an adaptation of the usual proof of differentiability with respect to parameters. We merely give an outline (see P. Lima [1] for details). Let $T_\lambda(x, \lambda) = \partial T(x, \lambda)/\partial\lambda$, $T_x(x, \lambda) = \partial T(x, \lambda)/\partial\lambda$ and consider the equation

$$z - T_x(x(\lambda), \lambda)z = T_\lambda(x(\lambda), \lambda)h$$

for given $h \in Y$. This equation has a unique solution $z(h, \lambda)$ which is linear in h. If $z(h, \lambda) = z(\lambda)h$, then one easily shows that $z(\lambda)$ is continuous in λ.

To complete the proof, one must show that $z(\lambda)$ satisfies the definition of the derivative of $x(\lambda)$; that is,

$$\omega \stackrel{\text{def}}{\equiv} x(\lambda + h) - x(\lambda) - z(\lambda)h = o(|h|)$$

as $|h| \longrightarrow 0$. An easy computation shows that $[I - T_x(x(\lambda), \lambda))]\omega - R(\omega, \lambda, h) = o(|h|)$ as $|h| \longrightarrow 0$ where $R(y, \lambda, h) = T(x + y, \lambda) - T(x, \lambda) - T_x(x, \lambda)y$ is $o(|y|)$ as $|y| \longrightarrow 0$. One now proves that this implies $\omega = o(|h|)$ as $|h| \longrightarrow 0$.

We illustrate the application of this theorem to prove the Hopf Bifurcation Theorem with respect to the delays for a differential difference equation. We impose more restrictions than in Hale [8] because the proof here will be based on the center manifold theorem. By showing there is an asymptotically stable local center manifold of dimension two which is continuously differentiable in λ, the problem is reduced to the usual one in ordinary differential equations.

Suppose $\Omega \subset \mathbf{R}^k$ is an open set (the parameter space), $C = C([-r, 0], \mathbf{R}^n)$, f: $\Omega \times C \longrightarrow \mathbf{R}^n$, L: $\Omega \times C \longrightarrow \mathbf{R}^n$ are continuous, $L(\alpha)\phi$ is linear in ϕ, $f(\alpha, \phi)$ has continuous first and second derivatives in ϕ, $f(\alpha, 0) = 0$, $\partial f(\alpha, 0)/\partial\alpha = 0$. With the notation $x_t(\theta) = x(t + \theta)$, $-r \leqslant \theta \leqslant 0$, consider the equation

$$\dot{x}(t) = L(\alpha)x_t + f(\alpha, x_t). \tag{10.4}$$

The first hypothesis is:

(H_1) *The matrix $\Delta(\alpha, \lambda) = \lambda I - L(\alpha)e^{\lambda \cdot}I$, where I is the identity matrix, is continuously differentiable in α, there is a pair of simple purely imaginary roots $\pm i\nu_0$, $\nu_0 > 0$, for $\alpha = \alpha_0$ and all other roots of the characteristic equation*

(10.5) $$\det \Delta(\alpha, \lambda) = 0$$

for $\alpha = \alpha_0$ *have negative real parts.*

Under hypothesis (H_1), there is a $\delta > 0$ and a simple characteristic root $\lambda(\alpha)$ which is C^1 in α for $|\alpha - \alpha_0| < \delta$ and $\lambda(\alpha_0) = i\nu_0$. Then $\operatorname{Re} \lambda(\alpha) = (\alpha - \alpha_0) \cdot \zeta(\alpha) + o(|\alpha - \alpha_0|)$ as $\alpha \longrightarrow \alpha_0$, where $\zeta(\alpha) \in \mathbf{R}^k$ is C^1 in α. Our next hypotheses are:

(H_2) $\zeta(\alpha_0) \neq 0$.

(H_3) *Considering* $L(\alpha)\phi$, $f(\alpha, \phi)$ *restricted to* $\mathbf{R}^k \times C^1([-r, 0], \mathbf{R}^n)$, *they have a derivative in* α *which is continuous in* α, ϕ *in the topology of* $\mathbf{R}^k \times C^1([-r, 0], \mathbf{R}^n)$.

For α a scalar, hypothesis (H_2) implies there are two eigenvalues crossing the imaginary axis at $\alpha = \alpha_0$. Hypothesis (H_3) does not imply that $L(\alpha)\phi$, $f(\alpha, \phi)$ have a derivative in α for $(\alpha, \phi) \in C$. For example, if $\alpha \in (0, r)$, the function $L(\alpha)\phi = \phi(-\alpha)$ satisfies hypothesis (H_3). This function clearly has no continuous derivative in α for $(\alpha, \phi) \in \mathbf{R} \times C$ since the linear operator $L(\alpha)$ is not continuous in α in the operator topology. In Hale [7], (H_3) was stated incorrectly. The hypothesis did not state that the continuity should be in α and ϕ.

THEOREM 10.2. *If* (H_1), (H_2), (H_3) *are satisfied, then there is an* $\epsilon > 0$ *such that for* $a \in \mathbf{R}$, $|a| < \epsilon$, *there is a* C^1*-manifold* $\Gamma_a \in \mathbf{R}^k$ *of codimension one,* Γ_a *is* C^1 *in* a, $\Gamma_0 = \{\alpha \in \mathbf{R}^k\colon \operatorname{Re} \lambda(\alpha) = 0, |\alpha - \alpha_0| < \epsilon\}$ *and, for every* $\alpha \in \Gamma_\alpha$, *there are a function* $\omega(\alpha, a)$, *an* $\omega(\alpha, a)$*-periodic function* $x^*(\alpha, a)$ *which is* C^1 *in* a *and* α, $\omega(\alpha_0, 0) = \omega_0 = 2\pi/\nu_0$, $x^*(\alpha, 0) = 0$ *and*

$$x^*(\alpha_0, 0)(t) = a\gamma \cos \nu_0 t + o(|a|) \quad \text{as } |a| \longrightarrow 0$$

where $\gamma \cos \nu_0 t$ *is a solution of* $\dot{x}(t) = L(\alpha_0)x_t$ *with* $\gamma \in \mathbf{R}^n$, $|\gamma| = 1$.

To indicate the proof, decompose the space C by the eigenvalues $(\lambda(\alpha), \bar{\lambda}(\alpha))$ as $C = P_\alpha \oplus Q_\alpha$ where P_α is two dimensional and spanned by the solutions of $\dot{x}(t) = L(\alpha)x_t$ corresponding to the eigenvalue $\lambda(\alpha)$, $\bar{\lambda}(\alpha)$. Let Φ_α be an $n \times 2$ matrix whose columns are a basis for P_α. If $T_\alpha(t)$ is the semigroup generated by this linear equation and $x_t = \Phi_\alpha y(t) + z_t$ with $y \in \mathbf{R}^2$, $z_t \in Q$, then equation (10.4) is equivalent to

$$\dot{y} = B_\alpha y + C_\alpha f(\alpha, \Phi_\alpha y + z_t),$$

$$z_t = T_\alpha(t)\psi + \int_0^t T_\alpha(t - s)\Psi f(\alpha, \Phi_\alpha y(s) + z_s)\,ds,$$

where B_α, C_α are two-by-two matrices C^1 in α, with the eigenvalues of B_α being $\lambda(\alpha)$, $\bar{\lambda}(\alpha)$, $\psi = z_0$, Ψ is the projection of the $n \times n$ matrix X_0 onto Q_α, $X_0(\theta) = 0$ for $\theta < 0$, $X_0(0) = I$ (see Hale [3] for details). There is an estimate $\|T_\alpha(t - s)|Q_\alpha|\| \leqslant Ke^{-\alpha t}$, $t \geqslant 0$, for some positive constants K, α.

To construct a local center manifold by means of simple looking formulas, let us suppose f and its derivative are bounded on $\Omega \times C$. If $\Gamma = C^1(\mathbf{R}^2, Q)$, $(y, h) \in \mathbf{R}^2 \times \Gamma$, let $\zeta(t, y, h)$, $\zeta(0, y, h) = y$, be the solution of the equation $\dot{\zeta} = B_\alpha \zeta + C_\alpha f(\alpha, \Phi_\alpha \zeta + h(\zeta))$

and define the operator

$$(10.6)\qquad K(\alpha, y)h = \int_{-\infty}^{0} T_\alpha(-s)\Psi f(\alpha, \Phi_\alpha \zeta(s, y, h)) + h(\zeta(s, y, h))\, ds.$$

A fixed point of $K(\alpha, h)$ is a center manifold of (10.4). An application of the Implicit Function Theorem yields a center manifold $M_\alpha = \{(y, \psi): \psi = h(\alpha, y)\}$ with the function $h(\alpha, y)$ continuous in α, y and having as many continuous derivatives in y as the function $f(\alpha, \phi)$ has in ϕ (finite, of course), $h(\alpha, 0) = 0$. The flow on the center manifold is determined by the ordinary equation

$$(10.7)\qquad \dot{y} = B_\alpha y + C_\alpha f(\alpha, \Phi_\alpha y(t) + h(\alpha, y(t)))$$

and is given by $(y(t), h(\alpha, y(t)))$. All solutions with initial values on M_α are defined for $t \in (-\infty, \infty)$. If we were considering only a local center manifold, we need here $t \in [-r, \beta)$, $\beta > 0$, which would be no loss in generality. If $y(t)$ is a solution of (10.7), this implies that $x_t = \Phi_\alpha y(t) + h(\alpha, y(t))$ is defined for $t \in (-\infty, \infty)$. Since x_t satisfies (10.4) and $\Phi_\alpha(\theta)$ is C^1 in θ this implies $h(\alpha, y(0))(\theta)$ is continuously differentiable in θ for $\theta \in [-r, 0]$. Thus, the fixed point set F_1 for $K(\alpha, y)$ must consist of continuously differentiable functions. Hypothesis (H_3) implies that condition (iv) of Lemma 10.1 is satisfied. Thus one can conclude that $h(\alpha, y)$ is C^1 in (α, y) (or C^k if all hypotheses are satisfied for derivatives up through order k) and that the vector field in (10.7) is C^1 in α, y. The proof of Theorem 10.2 is completed by applying the corresponding result for ordinary equations.

Some very interesting bifurcations occur in problems with several delays (see Hale [8], Nussbaum [1]).

Marsden and McCracken [1, p. 255] have a version of the Hopf Bifurcation Theorem for general semigroups of transformations satisfying some smoothness properties. Using Lemma 10.1, it should be possible to improve those results.

References

A. A. Andronov and F. A. Leontovich [1], *Sur la théorie de la variation de la structure qualitative de la division du plan en trajectories,* Dokl. Akad. Nauk SSSR **21** (1938), 427–430.

A. A. Andronov, F. A. Leontovich, I. I. Gordon and A. G. Meyer [1], *Theory of bifurcations of dynamical systems on a plane,* Wiley, New York, 1973.

A. A. Andronov and L. Pontryagin [1], *Systemes grossiers,* Dokl. Akad. Nauk SSSR **14** (1937), 247–251.

A. A. Andronov, A. A. Vitt and S. E. Khaikin [1], *Theory of oscillations,* Pergamon Press, New York, 1966.

D. V. Anosov [1], *Roughness of geodesic flows of compact Riemannian manifolds of negative curvature,* Soviet Math. Dokl. **3** (1962), 1068–1070.

——— [2], *Geodesic flows on compact Riemannian manifolds with negative curvature,* Proc. Steklov Inst. Math. **90** (1967); English transl., Amer. Math. Soc. Transl. (1969).

V. I. Arnol′d [1], *Additional chapters on the theory of ordinary differential equations,* Moskva "Nauka", 1978. (Russian)

J. Ball [1], *Saddle point analysis for an ordinary differential equation in a Banach space, and an application to dynamic buckling of a beam,* Nonlinear Elasticity, Academic Press, New York, 1973, pp. 93–160.

C. Bardos, H. Matano and J. Smoller [1], *Some results on the instability of the solutions of reaction diffusion equations* (preprint).

S. R. Bernfeld and L. Salvadori [1], *Generalized Hopf bifurcation and h-asymptotic stability,* J. Nonlinear Ana. **4** (1980), 1091–1108.

Yu. N. Bibikov [1], *Local theory of nonlinear analytic ordinary differential equations,* Lecture Notes in Math., vol. 702, Springer-Verlag, Berlin and New York, 1979.

G. D. Birkhoff [1], *Nouvelles recherches sur les systèmes dynamiques,* Mem. Pont. Acad. Sci. Novi Lyncaei **1** (1935), 85–216.

A. D. Br′juno [1], *Analytic form of differential equations,* Trans. Moscow Math. Soc. **25** (1971), 131–288; **26** (1972), 199–239.

P. Brunovsky and S. N. Chow [1], *Generic properties of stationary states of reaction-diffusion equations*, J. Differential Equations (to appear).

J. Carr [1], *Applications of centre manifold theory,* Lecture Notes LCDS LN 79–1, Lefschetz Center for Dynamical Systems, Div. Appl. Math., Brown University, Providence, R. I., 1979.

M. Cartwright [1], *Forced oscillations in nonlinear systems,* Theory Contr. Theory Nonlin. Osc. **1** (1950), 149–241; Ann. Math. Studies, no. 20, Princeton Univ. Press, Princeton, N. J., 1950.

—— [2], *Almost periodic flows and solutions of differential equations,* Proc. London Math. Soc. (3) **17** (1967), 355–380; Corrigenda (3) **17** (1967), 768.

—— [3], *Almost periodic differential equations and almost periodic flows,* J. Differential Equations **5** (1969), 167–181.

R. G. Casten and C. J. Holland [1], *Instability results for a reaction diffusion equation with Neumann boundary conditions,* J. Differential Equations **27** (1978), 266–273.

L. Cesari [1], *The alternative method in nonlinear oscillations,* Studies in Differential Equations (J. K. Hale, ed.), Studies in Math., vol. 14, Math. Assoc. Amer., New York, 1977.

N. Chafee [1[, *Generalized Hopf bifurcation and perturbation in full neighborhood of a given vector field,* Indiana Univ. Math. J. **27** (1978), 173–194.

—— [2], *Asymptotic behavior for solutions of a one dimensional parabolic equation with homogeneous Neumann boundary conditions,* J. Differential Equations **18** (1975) 111–134.

N. Chafee and E. F. Infante [1], *A bifurcation problem for a nonlinear partial differential equation of parabolic type,* J. Appl. Anal. **4** (1974), 17–37.

S. N. Chow and J. K. Hale [1], *Methods of bifurcation theory,* Springer-Verlag, Berlin and New York (to appear).

S. N. Chow, J. K. Hale and J. Mallet-Paret [1], *Applications of generic bifurcation.* I, II, Arch. Rational Mech. Anal. **59** (1975), 159–188; **62** (1976), 209–236.

—— [2], *An example of bifurcation to homoclinic orbits,* J. Differential Equations **37** (1980), 351–373.

S. N. Chow and H. O. Walther [1], *Chaos in delay differential equations* (to appear).

R. C. Churchill, G. Pecelli and D. L. Rod [1], *Stability transitions for periodic orbits in Hamiltonian systems,* Arch. Rational Mech. Anal. **73** (1980), 313–347.

E. A. Coddington and N. Levinson [1], *Theory of ordinary differential equations,* McGraw-Hill, New York, 1955.

C. Conley [1], *Isolated invariant sets and the Morse index,* NSF-CBMS Regional Conf. Series, no. 38, Amer. Math. Soc., Providence, R. I., 1978.

G. D. Cooperman [1], *α-condensing maps and dissipative systems,* Ph. D. Thesis, Brown University, 1978.

C. Corduneanu and V. Laksmikantham [1], *Equations with unbounded delay: A survey,* Tech. Rpt. No. 113, Univ. Texas, Arlington, Texas, 1979.

C. M. Dafermos and J. A. Nohel [1], *Energy methods for nonlinear hyperbolic Volterra integrodifferential equations,* Comm. Partial Differential Equations **4** (1979), 219–278.

J. C. deOliveira [1], *The generic C^i-property for a class of NFDEs,* J. Differential Equations **31** (1979), 329–336.

J. C. deOliveira and J. K. Hale [1], *Dynamic behavior from bifurcation equations,* Tohôku Math. J. (2) **32** (1980), 189–199.

S. D. Diliberto [1], *Perturbation theorems for periodic surfaces.* I, II, Rend. Circ. Mat. Palermo (2) **9** (1960), 265–299; **10** (1961), 111.

J. R. Dorroh and J. E. Marsden [1], *Smoothness of nonlinear semigroups* (preprint).

P. Fife [1], *Mathematical aspects of reaction diffusion systems,* Lecture Notes in Biomath., vol. 28, Springer-Verlag, Berlin and New York, 1979.

N. Fenichel [1], *The orbit structure of the Hopf bifurcation problem,* J. Differential Equations **17** (1975), 308–328.

W. F. Fitzgibbon [1], *Semilinear functional differential equations in Banach spaces,* J. Differential Equations **29** (1978), 1–14.

D. Flockerzi [1], *Existence of small periodic solutions of ordinary differential equations in* $\mathbf{R}^n$, Math. Inst. Univ. Wurzburg, (Aug. 1979).

J. Franks [1], *Differentiable Ω-stable diffeomorphisms,* Topology **11** (1972), 107–113.

A. Friedman [1], *Partial differential equations,* Holt, Rinehart and Winston, New York, 1969.

M. Golubitsky and W. F. Langford [1], *Classification and unfoldings of degenerate Hopf bifurcations,* J. Differential Equations (to appear).

D. Grobman [1], *Homeomorphisms of systems of differential equations,* Dokl. Akad. Nauk SSSR **129** (1959), 880–881.

J. Guckenheimer [1], *A brief introduction to dynamical systems,* Nonlinear Oscillations in Biology, Lectures in Appl. Math., vol. 17, Amer. Math. Soc., Providence, R. I., 1979, pp. 187–253.

—— [2], *Absolutely Ω-stable diffeomorphisms,* Topology **11** (1972), 196–197.

—— [3], *On a codimension two bifurcation* (preprint).

—— [4], *On quasiperiodic flow with three independent frequencies* (preprint).

J. K. Hale [1], *Ordinary differential equations,* Wiley, New York, 1969; 2nd ed., Krieger, 1980.

—— [2], *Stability from the bifurcation function,* Differential Equations (Ahmad, Keener, Lazer, eds.), Academic Press, New York, 1980, pp. 23–30.

—— [3], *Theory of functional differential equations,* Applied Math. Sci., vol. 3, 2nd ed., Springer-Verlag, Berlin and New York, 1977.

—— [4], *Retarded equations with infinite delays,* Functional Differential Equations and Approximation of Fixed Points (Peitgen and Walther, eds.), Lecture Notes in Math., vol. 730, Springer-Verlag, Berlin and New York, 1979, pp. 157–193.

—— [5], *Dynamical systems and stability,* J. Math. Anal. Appl. **26** (1969), 39–59.

—— [6], *Some recent results on dissipative systems,* Functional Differential Equations and Bifurcation (Izé, ed.), Lecture Notes in Math., vol. 799, Springer-Verlag, Berlin, Heidelberg and New York, 1980, pp. 152–172.

J. K. Hale [7], *Generic properties of an integro-differential equation,* Amer. J. Math. (to appear).

—— [8], *Nonlinear oscillations in equations with delays,* Nonlinear Oscillations in Biology (F. Hoppensteadt, ed.), Lectures in Appl. Math., vol. 17, Amer. Math. Soc., Providence, R. I., 1979, pp. 157–185.

—— [9], *Bifurcation near families of solutions,* Differential Equations, Acta Univ. Uppsal., Uppsala, 1977, pp. 91–100.

J. K. Hale and J. C. deOliveira [1], *Hopf bifurcation for functional equations,* J. Math. Anal. Appl. **74** (1980), 41–59.

J. K. Hale and J. Kato [1], *Phase space for retarded equations with infinite delays,* Funkcial. Ekvac. **21** (1978), 11–41.

J. K. Hale and P. Massatt [1], *Asymptotic behavior of gradient-like systems,* Proc. Conf. Nonlinear Anal. and Differential Equations, Gainesville, Fla., 1981 (to appear).

J. K. Hale and P. Táboas [1], *Interaction of damping and forcing in a second order equation,* Nonlinear Anal. **2** (1978), 77–84.

J. K. Hale and J. Vegas [1], *A nonlinear parabolic equation with varying domain,* Arch. Rational Mech. Anal. (to appear).

P. Hartman [1], *On the local linearization of differential equations,* Proc. Amer. Math. Soc. **14** (1963), 568–573.

—— [2], *Ordinary differential equations,* Wiley, New York, 1964.

—— [3], *A lemma in the theory of structural stability of differential equations,* Proc. Amer. Math. Soc. **11** (1960), 610–620.

J. Henrard [1], *On a perturbation theory using Lie transforms,* Celestial Mech. **3** (1970), 107.

D. Henry [1], *Geometric theory of semilinear parabolic equations,* Lecture Notes in Math., vol. 840, Springer-Verlag, Berlin and New York, 1981.

—— [2], *Gradient flows defined by parabolic equations,* Nonlinear Diffusion (Fitzgibbon and Walker, eds.), Pitman, New York, 1977, pp. 122–128.

P. Holmes [1], *Averaging and chaotic motions in forced oscillations,* SIAM J. Appl. Math. **38** (1980), 65–80.

—— [2], *Unfolding a degenerate nonlinear oscillator,* Nonlinear Dynamics (Helleman, ed.), New York Acad. Sci., 1980, pp. 473–488.

P. Holmes and J. Marsden [1], *A partial differential equation with infinitely many periodic orbits: Chaotic oscillations of a forced beam,* Arch. Rational Mech. Anal. (to appear).

L. N. Howard [1], *Nonlinear oscillations,* Nonlinear Oscillations in Biology (F. Hoppensteadt, ed.), Lectures in Appl. Math., vol. 17, Amer. Math. Soc., Providence, R. I., 1979, pp. 1–68.

L. N. Howard and N. Kopell [1], *Bifurcations and trajectories joining critical points,* Adv. in Math. **18** (1976), 306–358.

E. F. Infante and J. A. Walker [1], *A stability investigation for an uncompressible fluid with fading memory,* Arch. Rational Mech. Anal. **72** (1980), 203–218.

G. Iooss and D. D. Joseph [1], *Elementary stability and bifurcation theory,* Springer-Verlag, New York, Heidelberg and Berlin, 1980.

A. Kelley [1], *The stable, center-stable, center, center-unstable and unstable manifolds,* J. Differential Equations **3** (1967), 546–570.

—— [2], *Stability of the center stable manifold,* J. Math. Anal. Appl. **18** (1967), 336–344.

H. Kielhöfer [1], *Generalized Hopf bifurcation in Hilbert space,* Math. Methods Appl. Sci. **1** (1979), 498–513.

I. Kupka [1], *Contribution á la théorie des champs génériques,* Contr. Differential Equations **2** (1963), 457–484.

J. Kurzweil [1], *Global solutions of functional differential equations,* Lecture Notes in Math., vol. 144, Springer-Verlag, Berlin and New York, 1970.

O. A. Ladyženskaya [1], *A dynamical system generated by the Navier-Stokes equation,* J. Soviet Math. **3** (1975), 458–479.

W. F. Langford [1], *Periodic and steady state mode interactions lead to tori,* SIAM J. Appl. Math. **37** (1979), 22–48.

R. Leggett [1], *Remarks on set contractions and condensing maps,* preprint.

—— [2], *A note on "locally α-contracting" linear operators,* Boll. Un. Mat. Ital. **12** (1975), 124–126.

F. A. Leontovich [1], *On the creation of limit cycles from a separatrix,* Dokl. Akad. Nauk SSSR **78** (1951), 641–644.

N. Levinson [1], *Transformation theory of nonlinear differential equations of the second order,* Ann. of Math. (2) **45** (1944), 723–737.

—— [2], *A second order equation with singular solutions,* Ann. of Math. (2) **50** (1949), 127–153.

A. M. Liapunov [1], *Probléme général de la stabilité du movement,* Princeton Univ. Press, Princeton, N. J., 1949.

—— [2], *Sur les figures d'equilibre peu differentes des ellipsodies d'une masse liquide homogène donnee d'un mouvement de rotation,* Zap. Akad. Nauk, St. Petersburg (1906).

P. Lima [1], *Hopf bifurcation in equations with infinite delays,* Ph. D. Thesis, Brown University, 1977.

J. Mallet-Paret [1], *Negatively invariant sets of compact maps and an extension of a theorem of Cartwright,* J. Differential Equations **22** (1976), 331–348.

—— [2], *Generic properties of retarded functional differential equations,* Bull. Amer. Math. Soc. **81** (1975), 750–752.

—— [3], *Generic periodic solutions of functional differential equations,* J. Differential Equations **25** (1977), 163–183.

R. Mañé [1], *Absolute and infinitesimal stability,* Sympos. Dynamical Systems (Warwick, 1974), Lecture Notes in Math., vol. 468, Springer-Verlag, Berlin and New York, 1975, pp. 24–26.

R. Mañé [2], *On the dimension of the compact invariant set of certain nonlinear maps*, preprint.

P. Manselli and K. Miller [1], *Dimensionality reduction methods for efficient numerical solutions, backward in time, of parabolic equations with variable coefficients*, SIAM J. Math. Anal. **11** (1980), 147–159.

F. Marchetti, P. Negrini, L. Salvadori and M. Scalia [1], *Liapunov's direct method in approaching bifurcation problems*, Ann. Mat. Pura Appl. (4) **108** (1976). 211–225.

J. E. Marsden [1], *Qualitative methods in bifurcation theory*, Bull. Amer. Math. Soc. **84** (1979), 1125–1148.

J. E. Marsden and M. McCracken [1], *The Hopf bifurcation and its applications*, Appl. Math. Sci., vol. 19, Springer-Verlag, Berlin and New York, 1976.

R. H. Martin, Jr. [1], *Nonlinear operators and differential equations in Banach spaces*, Wiley, New York, 1976.

P. Massatt [1], *Some properties of condensing maps*, Ann. Mat. Pura Appl. (to appear).

—— [2], *Stability and fixed points of dissipative systems*, J. Differential Equations **40** (1981) (to appear).

—— [3], *Linear β-condensing maps are α-contractions* (preprint).

—— [4], *Asymptotic behavior of a strongly damped nonlinear wave equation* (preprint).

—— [5], *Attractivity properties of α-contractions* (preprint).

H. Matano [1], *Asymptotic behavior and stability of solutions of semilinear diffusion equations*, Publ. Res. Inst. Math. Sci. **15** (1979), 401–454.

—— [2], *Convergence of solutions of one-dimensional semilinear parabolic equations*, J. Math. Kyoto Univ. **18** (1978), 221–227.

V. K. Mel'nikov [1], *On the stability of the center for time periodic solutions*, Trans. Moscow Math. Soc. **12** (1973), 3–56.

K. Meyer and D. S. Schmidt [1], *Entrainment domains*, Funkialoj Ekvacioj **20** (1977), 171–192.

P. Michor [1], *The division theorem on Banach spaces*, Math. Inst. Univ. Wien, Austria, preprint.

K. Miller [1], *Nonunique continuation for certain ODE's in Hilbert space and for uniformly parabolic and elliptic equations in self-adjoint divergence form*, Lecture Notes in Math., vol. 316, Springer-Verlag, Berlin and New York, 1973, pp. 85–101.

N. Minorsky [1], *Nonlinear oscillations*, Van Nostrand, Princeton, N. J., 1962.

J. Moser [1], *On a theorem of Anosov*, J. Differential Equations **5** (1969), 411–440.

—— [2], *Stable and random motions in dynamical systems*, Princeton Univ. Press, Princeton, N. J., 1973.

P. Negrini and L. Salvadori [1], *Attractivity and Hopf bifurcation*, Nonlinear Anal. **3** (1979), 87–99.

S. Newhouse [1], *Lectures on dynamical systems*, Dynamical Systems (Guckenheimer, Moser, Newhouse, eds.), Birkhäuser, Boston, 1980, pp. 1–115.

S. Newhouse [2], *On simple arcs between structurally stable flows,* Sympos. Dynamical Systems (Warwick, 1974), Lecture Notes in Math., vol. 468, Springer-Verlag, Berlin and New York, 1975, pp. 209–233.

S. Newhouse and M. M. Peixoto [1], *There is a simple arc joining any two Morse-Smale flows,* Asterique **31** (1976), 15–42.

Z. Nitecki [1], *Differentiable dynamics,* M.I.T. Press, Cambridge, Mass., 1971.

R. Nussbaum [1], *Periodic solutions of analytic functional differential equations are analytic,* Michigan Math. J. **20** (1973), 249–255.

—— [2], *Periodic solutions of nonlinear autonomous functional differential equations,* Functional Differential Equations and Approximation of Fixed Points (Peitgen and Walther, eds.), Lecture Notes in Math., vol. 730, Springer-Verlag, Berlin and New York, 1979, pp. 283–325.

W. M. Oliva [1], *The behavior at infinity and the set of global solutions of retarded functional differential equations,* Sympos. Functional Differential Equations, (São Carlos, Brasil, 1975), Colecao ATAS, vol. 8, Soc. Brasileira de Mat., 1975.

J. Palis [1], *On Morse-Smale dynamical systems,* Topology **8** (1969), 385–404.

—— [2], *A note on Ω-stability,* Global Analysis, Proc. Sympos. Pure Math., vol. 14, Amer. Math. Soc., Providence, R. I., 1970, pp. 221–222.

J. Palis and S. Smale [1], *Structural stability theorems,* Global Analysis, Proc. Sympos. Pure Math., vol. 14, Amer. Math. Soc., Providence, R. I., 1970, pp. 223–232.

J. Palis and W. deMelo [1], *Introducão aos sistemas dinamicos,* Inst. Math. Pura Aplicada, Rio de Janeiro, 1978.

A. Pazy [1], *Semigroups of linear operators and applications to partial differential equations,* Lecture Nôtes, no. 10, Univ. Maryland, College Park, Maryland, 1974.

M. Peixoto [1], *Structural stability on two-dimensional manifolds,* Topology **1** (1972), 101–120.

—— [2], *Generic properties of ordinary differential equations,* Studies in Differential Equations (J. K. Hale, ed.), Studies in Math., vol. 14, Math. Assoc. Amer., New York, 1977, pp. 52–92.

—— [3], *Dynamical systems,* Academic Press, New York, 1973.

—— [4], *On structural stability,* Ann. of Math. (2) **69** (1959), 189–222.

—— [5], *On the classification of flows on 2-manifolds,* Dynamical Systems, Academic Press, New York, 1973, pp. 380–420.

V. A. Pliss [1], *A reduction principle in the theory of stability of motion,* Izv. Akad. Nauk SSSR Ser. Mat. **28** (1964), 1297–1324.

H. Poincaré [1], *Les méthodes nouvelles de la mécanique céleste,* vol. 3, Gauthier-Villars, Paris, 1892.

—— [2], *Sur l'equilibre d'une masse fluids animes d'un mouvement de rotation,* Acta Math. **7** (1885), 259–380.

C. Robinson [1], *Structural stability of C^1-diffeomorphisms,* J. Differential Equations **22** (1976), 28–73.

B. N. Sadovskii [**1**], *Limit compact and condensing operators,* Uspehi Mat. Nauk **27** (**163**) (1972), 81–147 = Russian Math. Surveys, 85–146.

K. Schumacher [**1**], *Existence and continuous dependence for functional differential equations with unbounded delay,* Arch. Rational Mech. Anal. **67** (1978), 315–335.

M. Shub [**1**], *Stabilité globale des systèmes dynamiques,* Astérique, 1978.

M. Slemrod [**1**], *A hereditary partial differential equation with applications in the theory of simple fluids,* Arch. Rational Mech. Anal. **62** (1976), 303–321.

——— [**2**], *Instability of steady shearing flows in a nonlinear viscoelastic fluid,* Arch. Rational Mech. Anal. **68** (1978), 211–225.

S. Smale [**1**], *Structurally stable systems are not dense,* Amer. J. Math. **88** (1966), 491–496.

——— [**2**], *Differentiable dynamical systems,* Bull. Amer. Math. Soc. **73** (1967), 747–817.

——— [**3**], *Stable manifolds for differential equations and diffeomorphisms,* Ann. Scuola Norm. Sup. Pisa **18** (1963), 97–116.

——— [**4**], *On gradient dynamical systems,* Ann. of Math. (2) **74** (1961), 199–206.

——— [**5**], *Diffeomorphisms with many periodic points,* Differential and Combinatorial Topology (S. S. Cairns, ed.), Princeton Univ. Press, Princeton, N. J., 1965, pp. 63–80.

J. Smoller and A. Wasserman [**1**], *Global bifurcation of steady state solutions,* J. Differential Equations **39** (1981), 269–290.

J. Sotomayor [**1**], *Structural stability and bifurcation theory,* Dynamical Systems (M. Peixoto, ed.), Academic Press, New York, 1973, pp. 549–560.

——— [**2**], *Generic one parameter families of vector fields,* Inst. Hautes Études Sci. Publ. Math. **43** (1973), 5–46.

F. Takens [**1**], *Unfolding of certain singularities of vector fields: Generalized Hopf bifurcations,* J. Differential Equations **14** (1973), 476–493.

——— [**2**], *Forced oscillations and bifurcations,* Applications of Global Analysis, Comm. 3 of the Math. Inst. Rikjsuniversitat Utrecht (1974).

M. A. Teixeira [**1**], *Generic bifurcation in manifolds with boundary,* J. Differential Equations **25** (1977), 65–89.

K. Uhlenbeck [**1**], *Generic properties of eigenfunctions,* Amer. J. Math. **98** (1976), 1059–1078.

M. Urabe [**1**], *Nonlinear autonomous oscillations,* Academic Press, New York, 1967.

A. Vanderbauwhede [**1**], *Local bifurcation theory and symmetry,* Habilitation, Ghent, 1980.

J. Vegas [**1**], *Bifurcation in semilinear Neumann problems under perturbation of the domain,* Ph. D. Thesis, Brown University, Providence, R. I., 1981.

H. O. Walther [**1**], *Dynamics of a model delay equation,* J. Nonlinear Ana. (to appear).

G. Webb [**1**], *Existence and asymptotic behavior for a strongly damped nonlinear wave equation* (preprint).

R. Williams [**1**], *The "DA" maps of Smale and structural stability,* Proc. Sympos. Pure Math., vol. 14, Amer. Math. Soc., Providence, R. I., 1970, pp. 329–334.

ABCDEFGHIJ–AMS–8987654321